THE DEFINITIVE GUIDE TO ORDER FULFILLMENT AND CUSTOMER SERVICE

PRINCIPLES AND STRATEGIES FOR PLANNING, ORGANIZING, AND MANAGING FULFILLMENT AND SERVICE OPERATIONS

Council of Supply Chain
Management Professionals
and
Stanley E. Fawcett
Amydee M. Fawcett

Vice President, Publisher: Tim Moore
Associate Publisher and Director of Marketing: Amy Neidlinger
Executive Editor: Jeanne Glasser Levine
Consulting Editor: Chad Autry
Operations Specialist: Jodi Kemper
Cover Designer: Chuti Prasertsith
Managing Editor: Kristy Hart
Senior Project Editor: Lori Lyons
Copy Editor: Karen Annett
Proofreader: Debbie Williams
Senior Indexer: Cheryl Lenser
Compositor: Nonie Ratcliff
Manufacturing Buyer: Dan Uhrig

© 2014 by Council of Supply Chain Management Professionals

Published by Pearson Education, Inc.
Upper Saddle River, New Jersey 07458

For information about buying this title in bulk quantities, or for special sales opportunities (which may include electronic versions; custom cover designs; and content particular to your business, training goals, marketing focus, or branding interests), please contact our corporate sales department at corpsales@pearsoned.com or (800) 382-3419.

For government sales inquiries, please contact governmentsales@pearsoned.com.

For questions about sales outside the U.S., please contact international@pearsoned.com.

Company and product names mentioned herein are the trademarks or registered trademarks of their respective owners.

All rights reserved. No part of this book may be reproduced, in any form or by any means, without permission in writing from the publisher.

Printed in the United States of America

2   17

ISBN-10: 0-13-345386-3
ISBN-13: 978-0-13-345386-7

Pearson Education LTD.
Pearson Education Australia PTY, Limited.
Pearson Education Singapore, Pte. Ltd.
Pearson Education Asia, Ltd.
Pearson Education Canada, Ltd.
Pearson Educación de Mexico, S.A. de C.V.
Pearson Education—Japan
Pearson Education Malaysia, Pte. Ltd.

Library of Congress Control Number: 2013953113

To our families—up and downstream.

Our parents (Stanley, Francine, Arthur Dean, and Amy) have provided years of encouragement.

Our children (Carisa, Tannen, Kjanela, Dallin, Keana, and Taft) bring constant inspiration to our lives.

Thank you!

CONTENTS

CHAPTER 1 Meeting Customers' Real Needs: The Nature
of Service System Design 1
    Meeting Customers' Real Needs 3
        Today's Dual Customer Challenge 3
    Creating Customer Value 7
        Cost 9
        Quality 9
        Delivery 10
        Responsiveness 11
        Innovation 12
        Total Order Performance—A Synergistic Approach 13
    Contributing to Customer Satisfaction 15
        Customer Service Strategies 17
        Customer Satisfaction Strategies 18
        Customer Success Strategies 21
    Service System Design 23
        Touch Points 24
        Orchestration 24
        Value Gaps 25
        Loyalty and Competitive Advantage 27
    Conclusion 29
    Endnotes 30

CHAPTER 2 Fulfilling Orders: The Nature of Modern Order Cycle Management 35

    Fulfilling Orders 38

    The Deliverables of an Order Fulfillment System 39

        Product Availability 40

        Timely Delivery.. 41

        Transparent, Reliable Service 43

        Service Recovery 43

        Efficient Operations 45

    The Details of an Order Fulfillment System 45

        Mapping the Order Delivery Cycle: The SCOR Model......... 46

        Providing Postsales Customer Service....................... 52

    The Cost of Order Fulfillment Failures 54

        The Cost of Stockouts...................................... 55

        The Cost of Supply Chain Glitches......................... 57

    Conclusion...58

    Endnotes..59

CHAPTER 3 Developing a Winning Customer Fulfillment Strategy...... 63

    Developing a Winning Customer Fulfillment Strategy............66

    Managing Customer Relationships for Profitable Growth.........66

    Managing Transactional Relationships 67

    Managing Strategic Alliances 69

        Phase 1: Internal Planning.................................. 72

        Phase 2: Collaborative Planning 72

        Phase 3: Day-to-Day Management.......................... 73

        Relationship Takeaways for Fulfillment Strategy Design 74

    Segmentation Tools and Techniques............................77

        ABC Classification 77

        Customer Profitability Analysis 82

    Tailored Logistics: The Right Service for Each
    Customer Segment 89
    Conclusion 91
    Endnotes 92

CHAPTER 4 Configuring the Network for Successful Fulfillment 95

    Configuring the Network for Successful Fulfillment 97
        The Nature of Network Configuration 99
    Systems Thinking and Order Fulfillment Configuration 115
    Global Implications for Network Configuration 118
        Compatibility 119
        Configuration 120
        Coordination 120
        Control 121
        Continuity 121
    Conclusion 123
    Endnotes 124

CHAPTER 5 Implementing an Enabling Technology Strategy 131

    Implementing an Enabling Technology Strategy 134
    The Nature of Information-Technology Enablement 136
        A Closer Look at Connectivity 138
        A Closer Look at Willingness 140
    Moving Toward Information Enablement 142
        Understanding Investment Patterns 142
        Following a Proven Path 145
    Pieces of the IT-Enablement Puzzle 149
        Customer Relationship Management Systems 155
        Order Processing Systems 159
    Conclusion 162
    Endnotes 164

CHAPTER 6 Assessing Performance for Success and Improvement 171

    Assessing Performance for Success and Improvement 174

    The Nature and Power of Performance Measurement 175

        Measurement Informs Understanding 176

        Measurement Motivates Behavior 176

        Measurement Drives Execution 177

    Measurement Practice—Understanding the Big Picture 178

        Holistic Process and Supply Chain Measurement 180

        Customer-Centric Measurement 187

        Balanced Scorecards 188

    Measurement Practice—Delving into the Details 194

        Product Availability 196

        Order Cycle Time 198

    Conclusion 201

    Endnotes 202

Index 207

ACKNOWLEDGMENTS

We acknowledge the contributions of colleagues across the supply chain community who have brought SCM from the back office to center stage as the value creation engine of the modern corporation—and our modern economy. We would also like to recognize the following three groups:

- The CSCMP team—Rick Blasgen, Kathleen Hedland, Ann Neumann, Kathy McInerney, Heather Morys, and Jessica D'Amico, who deserve acknowledgment for providing vision and indefatigable efforts to advance the discipline.

- The SCPro Committee—Dr. Ted Stank (University of Tennessee), Dr. Chris Moberg (Ohio University), Dr. Tom Speh (Miami University), and Dr. Brian Gibson (Auburn University) deserve recognition for cultivating the SCPro concept.

- The book series editors—Dr. Chad Autry (University of Tennessee) and Jeanne Glasser Levine at Pearson deserve credit for creating the supply chain book series and guiding the development process.

ABOUT THE AUTHORS

Stanley E. Fawcett is the John B. Goddard Endowed Chair in Global Supply Chain Management at Weber State University. Stan taught at Michigan State University and Brigham Young University before joining Weber State. Stan is an innovative, award-winning teacher who has taught academic and executive programs in Asia, Europe, and North and South America. He has published more than 130 articles and six books on supply chain topics. His research has recently appeared in the following leading journals: *Decision Sciences Journal, Journal of Business Logistics, Journal of Finance, Journal of Small Business Management*, and *Journal of Supply Chain Management*. Stan is the coeditor-in-chief of the *Journal of Business Logistics*. Stan's two core philosophies are mentoring and collaboration. His greatest satisfaction as an academic comes from helping colleagues and students achieve higher levels of personal and professional success.

Amydee M. Fawcett, PhD, is Assistant Professor of SCM in the John B. Goddard School of Business and Economics at Weber State University. Dr. Fawcett's research explores the dynamics of effective collaboration in a variety of settings, including collaborative planning, forecasting, and replenishment (CPFR) and humanitarian assistance and disaster relief (HADR). She has coauthored ten articles, including two best papers: the E. Grosvenor Plowman Best Paper Award from CSCMP Educator's Conference and the Hal E. Fearon Best Paper Award from *Journal of Supply Chain Management*. Her teaching focuses on supply chain relational dynamics.

1

MEETING CUSTOMERS' REAL NEEDS: THE NATURE OF SERVICE SYSTEM DESIGN

Opening Story: The High-Service Sponge

September 26

Diane Clair, director of logistics at Dynamic World Corp. (DWC), a leading consumer packaged goods (CPG) company, had been running nonstop all morning. Now, as she sat down to put the finishing touches on the logistics team's five-year technology plan, she was running behind. Tomorrow morning, she would be arguing for some big-dollar investments in new demand management planning systems. No sooner had she clicked open her PowerPoint, than she heard an agitated knock at her door. Doug Hassle, DWC's North America marketing VP, stood there—and he wasn't smiling. Diane responded, "Good morning, Doug. Come in and sit down." As Doug entered, he said, "Diane, the wheels just came off. Deb Gale, GMM over at Monster, Inc., just called. She was ticked. Your team missed a delivery window at their Denver cross-dock facility. Worse, this is twice in one week we have failed to deliver as promised. Monster is our largest, most demanding customer. Deb made sure I remembered that little detail. She didn't hesitate to share her feelings about our recent fulfillment failures."

"Sounds like you've had a tough morning, Doug. Sorry about that brutal call. Let's find out what happened," Diane said as she picked up her phone. She dialed David England, a senior transportation manager, to find out what happened. David responded quickly, "The shipment left our Distribution Center on time. Our tracking system says the truck should arrive in about an hour. I'll make sure it does and shoot you a text when it is docked." David hesitated, and then expressed disbelief at Deb Gale's negative perceptions of DWC's delivery record, noting, "I can't imagine why Deb Gale is so upset. We're an industry leader. Our on-time delivery and complete orders performance is outstanding. We've never performed better."

Confident David would resolve the immediate crisis, Diane hung up. Doug, however, wasn't placated, and said as much, "Diane, for someone who just dropped the ball with our most important account, David sounded a bit overconfident. He might not really grasp the situation." Inwardly, Diane scowled. She trusted David. "Well, let's double-check and see how we are really performing." Diane picked up the phone again, this time calling Paul Osterhaus, vice president of information technology, to verify DWC's delivery performance. With just a few clicks of the mouse, Paul pulled up the key stats, confirming that DWC had dramatically improved its on-time delivery over the past year. He said, "You guys have really done a nice job. You've bumped your performance from 95 percent to 98 percent on-time delivery over the past 12 months. And you're shipping 97 percent complete orders. It seems you're hitting on all cylinders and achieving best-in-class standards." Paul added, "With service levels looking so good, you should be able to justify those new IT investments."

Doug still wasn't convinced, saying, "I'm sure you like what you're hearing, but those stats don't change the fact that Deb Gale just chewed me out. We're clearly not delivering to Monster's expectations. And though the chargebacks for late deliveries make these failures expensive, the real cost is relational. We can't afford for Monster to drop us as a supplier or to reduce the number of facings they allot us. Deb drove this point home, saying, 'You can't afford not to meet our needs.' She's right! Our other key accounts are just as demanding. If we are dropping the ball with Monster, we are likely disappointing the others as well. Come January, Monster is tightening its delivery time windows and Deb has informed me that they will expect us to take on more value-added services." As Doug stood up to leave, he added, "By the way, they are lengthening payment terms—effectively paying us less for what we do. I hope your team can raise the performance bar." Diane acknowledged that DWC needed to step up service even higher, concluding, "You're right, today's market is a tough place to do business. Our best customers are more than happy to soak up every ounce of service we can provide—and then they squeeze us a little more."

Still unsettled after Doug had left her office, Diane called David to begin a new conversation on DWC's customer fulfillment capabilities. After her initial greeting, Diane said, "David, despite all the customer-oriented initiatives we have pursued over the past two years, we dropped the ball today—and not just any ball." She smiled as she continued, "If we are going to drop a ball, let's make sure it doesn't belong to Monster. Today's experience reiterates our need to rethink our service strategy. Not all customers are created equal; yet, we still measure and manage to averages. Across-the-board excellence is a great goal, but maybe our one-size-fits-all approach is outdated. We just don't have the resources to be perfect all the time. So, what are we going to do about it? What service experience should we promise? What infrastructure do we need to put in place to deliver to promise? David, this is a big deal. I don't want to have this same conversation with Doug again. I need you to put a team together and get this figured out."

Consider as you read:

1. Why should logistics managers worry about customer service?
2. Keeping your answer from Question 1 in mind, what is the relationship between customer expectations, a firm's service capabilities, and ultimate satisfaction?
3. What questions would you include on a customer satisfaction checklist to make sure you had a comprehensive, well-thought-out customer fulfillment strategy in place?

Meeting Customers' Real Needs

> "There is only one valid definition of business purpose: to create a customer. What the customer buys and considers value is never just a product. It is always a utility, that is, what a product or service does for him."
> —Peter Drucker[1]

Why do companies exist? You might argue a variety of valid answers, but you can't ignore Peter Drucker's advice to focus on the customer. Without customers, your company cannot earn a profit, create jobs, or be a responsible corporate citizen. Creating customer value must be at the core of your business model design as well as your day-to-day decision making. Ultimately, if you want your company to survive in today's tremendously competitive global marketplace, you must meet customers' needs as well as or better than the competition.

Today's Dual Customer Challenge

Your management team must therefore develop an intimate understanding of customer needs. This fact brings two vital questions to the forefront:

- Who are your customers?
- How well do you know your customers and their needs?

The answers might not be as obvious as you think. Recognizing that companies almost always serve a host of distinct customers, your initial response to these questions is likely, "We serve a variety of channels and a diverse set of customer segments. Our knowledge of their needs varies based on how important we have defined each customer, segment, or channel to be." The result: You know some customers very well. Others leave you guessing.

As you think more deeply about the two questions above, you might consider the tiers of customers that compose your supply chain. Unless your company is a retailer, the goods and services you sell are used to help your customers deliver value to their customers. Depending on your supply chain positioning (you probably participate in multiple, distinct supply chains), you might have two, three, or more tiers of downstream customers. Undoubtedly, you spend most of your time and effort working to meet the needs of your company's immediate, first-tier customers. Customers further downstream may only be an abstraction in your planning scenarios. This reality can be dangerous as your long-term success increasingly depends on how well you grasp the needs of these distant downstream customers.

Seeing Downstream

Figure 1-1 portrays two approaches to understanding customer needs. Panel A depicts a traditional approach. Each company focuses on understanding customer expectations and market imperatives at the first-tier level. Consumer packaged goods companies are the exception. They have always paid attention to their retail customers as well as the end consumer who uses their products. Panel B depicts a more holistic approach that focuses members of each supply chain tier on the end customer. The logic is that supply chains compete against other supply chains in today's global marketplace.[2] In theory, to maximize value creation—and thus meet the needs of the end customer better than rival supply chains—every member of the supply chain team should understand the needs and wants of the end customer.

In practice, few supply chains achieve the holistic customer understanding shown in Panel B. However, some companies are working diligently toward this ideal. They proactively seek to help their suppliers learn more about downstream customers. Walmart, for instance, knows that if its suppliers don't deliver, it won't either. Walmart depends on suppliers to provide access to the hot products that customers desire—and it needs them to deliver on time. To help keep suppliers on the same page, Walmart employs a Web-based tool called Retail Link. Supplier account managers log in to find out how well their products are selling. They can see daily inventory levels and flow-through rates. They can also track margins. More impressively, they can do this at the individual store level or in different geographic regions. Suppliers can, thus, strive to match supply to demand—both in terms of the types of products to sell at individual stores and in terms of when and how much to ship to keep shelves stocked.

Walmart's efforts to help suppliers understand customer needs do not end with Retail Link. Walmart uses its channel position and day-to-day interaction with customers to become a sensor for its entire supply chain. As Walmart entered the China retail market, store managers noticed that laundry detergent was not selling—at least not in the expected quantities. Upon investigation, Walmart discovered that few Chinese households owned washing machines. Walmart turned to Procter & Gamble (P&G), asking it to develop a new detergent specially designed for hand washing. Tide White's launch was a huge hit, selling 35,000 units in 15 stores in two weeks. Acting as P&G's eyes and ears, Walmart spotted an unmet need. Working together kept Walmart's shelves stocked with the products its customers want while improving P&G's research and development effectiveness.[3] The two are partners in profit. Moreover, the entire supply chain wins when these collaborative efforts bring more customers to Walmart's stores.

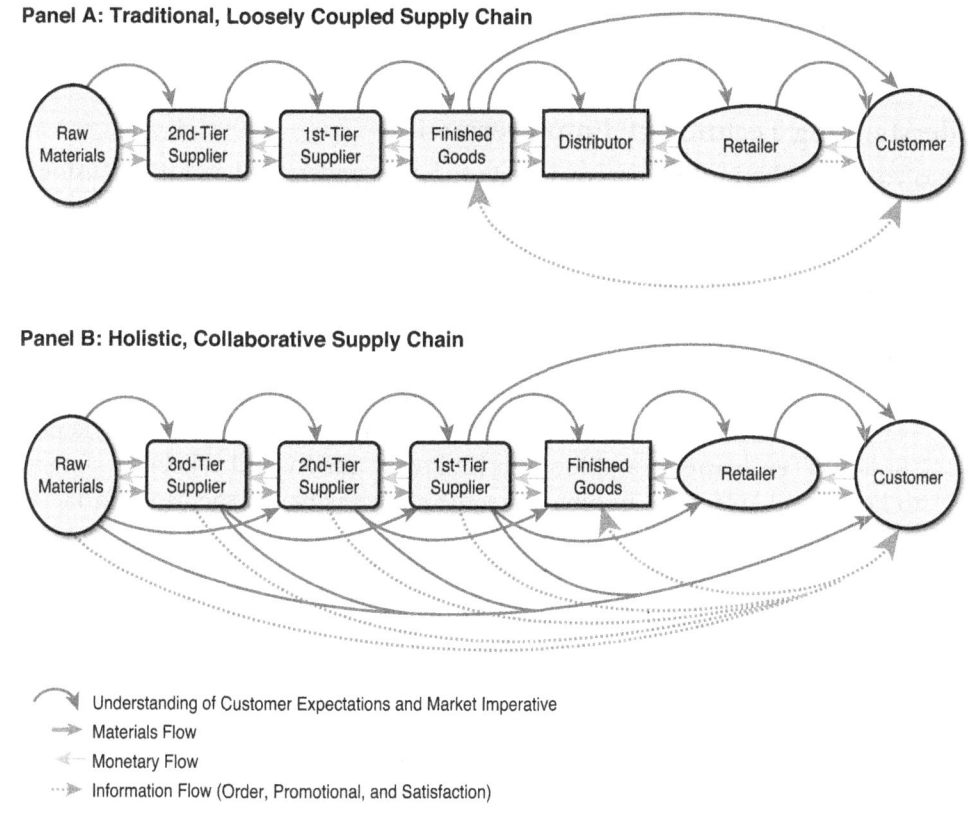

Figure 1-1　Understanding customer needs[4]

You should take the following three learning points away from this discussion:

- If your company does not meet the needs of your immediate customers, they will buy from someone else—a supplier that helps them succeed. If P&G had been unwilling or unable to develop a new laundry detergent, Walmart would have certainly turned to one of P&G's rivals—perhaps Unilever. This process is called disintermediation.

- The end customer takes center stage in well-designed and well-managed supply chains. The end customer is the only customer who really puts money into the chain. Everyone else simply recycles it. Every organization in a supply chain should seek to improve its ability to contribute to satisfying the end customer's needs and wants.

- Companies win when they participate as members of a winning supply chain team. If you don't understand downstream competitive dynamics, including customer requirements, you will find it difficult to allocate the resources needed to be a supplier of choice in these winning supply chains.

Understanding Technology Empowerment

Even as you are gathering and analyzing data to better understand and meet customer needs, your customers are using modern technology systems to meet their own needs. They know that information is power and they use it to get a better deal on the things they are buying. Consider the following four Business to Consumer (B2C) examples to see how the Internet is a game changer:

- **Information acquisition**—Today's savvy shopper compiles product specifications and compares prices without leaving the comfort of home. A car buyer, for example, can check out Car and Driver write-ups, consider Consumer Reports "Best Buy" evaluations, and look up JD Power quality and reliability statistics to help decide what car to buy. Once the choice is made, she can find factory invoice information on Edmunds.com. As she enters the showroom, she is an information-empowered consumer, ready and confident to engage in aggressive negotiations.

- **The buying experience**—Today's smart shopper wants a real experience, but will go virtual to save money. Showrooming exemplifies this behavior. Showrooming is when a shopper physically visits a store to check out a product, but then purchases the product from a lower-priced online retailer. Best Buy, a leading—and, until recently, very profitable—bricks-and-mortar electronics retailer, has struggled to compete in today's showrooming environment. By using their rivals' physical stores as their showrooms, Amazon.com and other online retailers lower their own cost structure. To compete, Best Buy must improve its own Internet presence even as it lowers prices and creates value that cannot be matched by clicks-only retailers.

- **Comparison shopping (on steroids)**—Today's sharp shopper employs mobile technology. By downloading "TheFind" (or a comparable app) to her iPhone, she can compare prices in real time—even for impulse buys in the middle of a shopping trip. TheFind advertises the following capability: "Find every product from every store, every coupon and every review. Everything you need when shopping to quickly decide what to buy and where to buy it."[5]

- **Postpurchase evaluation**—Today's shrewd shopper knows that technology empowerment extends beyond the purchase. At Amazon.com, shoppers can rate every product and every transaction. They can also turn to blogs and video-posting sites to voice their horror or delight to anyone, anywhere in the world who has access to a computer. For example, when United Airlines ramp workers broke Dave Carroll's guitar, he spent a year—to no avail—trying to get United to pay for the repairs. Frustrated, he told Ms. Erlweg, United's customer service representative, that he would post three videos to the Internet. To United's dismay, his first video, "United Breaks Guitars," attracted one million hits in three days (3.5 million hits in ten days). Modern customers can creatively share their complaints.

In the Business to Business (B2B) setting, sourcing professionals can use technology to do many of the same things. They can use ThomasNet (and other online registers) to identify potential suppliers, evaluate capabilities, and compare pricing.[6] They can tap into data analytics to track and benchmark supplier performance. They can employ e-sourcing events (sometimes called reverse auctions) to place rival suppliers in a real-time competitive-bidding environment. Just like their consumer contemporaries, supply managers are leveraging technology to shift channel power in their favor.

To summarize, customer access to viable options through competing supply chains combined with technology empowerment have changed the competitive rules for most companies. Companies must relentlessly deliver higher levels of service at lower costs. Their best customers, who behave as "high-service sponges," will no longer settle for *average* service. To fuel their own quest for market success, these high-service sponges seek and "soak up" their suppliers' resources. What does this mean for you? You need to understand how your customers define **value** and determine **satisfaction**.

Creating Customer Value

Why do you need an accurate understanding of *how* your customers define value? First, customers make purchase decisions based on the value they expect to obtain. The question in the mind of decision makers is, "Does the value justify the cost?" Second, customers employ similar criteria to evaluate supplier performance as they used to select suppliers in the first place. If your quest is to drive revenue growth—that is, obtain and retain customers—you need to deeply understand what customers value.

Economists often discuss value in terms of utilities. As Peter Drucker noted, utility is what a "product or service does" for the customer. Four core utilities are often discussed:

- **Form** utility is the primary responsibility of purchasing and operations managers who acquire inputs and *transform* them into products or services of greater customer value.

- **Possession** utility falls within marketing's domain and consists of efforts to communicate (i.e., promote) a product's value and then facilitate the exchange process.

- **Time** utility emerges from effective management of all value-added processes that influence *when* a product is available for purchase. Logistics managers make the inventory and transportation decisions that ultimately determine availability's *time* dimension.

- **Place** utility is primarily the charge of supply chain managers who ensure that products and services are *where* customers expect to find them—when they are needed.[7]

A correct understanding of economic utilities informs the design of a company's supply network and value-added processes; however, form, possession, time, and place utilities are seldom explicitly discussed by decision makers in the supplier selection and evaluation processes.

To gain direct insight into how your customers define value, you might want to gather supplier scorecards from a sample of your customers. You will find five criteria are common across the majority of the scorecards: *cost, quality, delivery, responsiveness* (also known as flexibility), and *innovation* (see Figure 1-2). Indeed, scholars have long argued that these value dimensions are the key to meeting customers' real needs.[8] To win tomorrow's competitive battles, you must grasp the nature of these value dimensions and build the systems to create and deliver them.

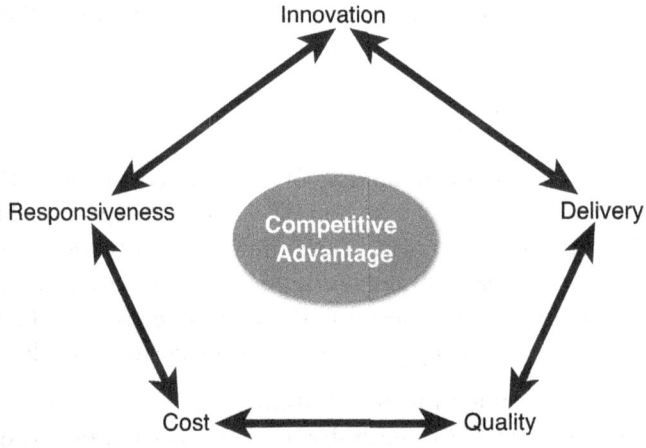

Figure 1-2 Dimensions of value creation

Cost

Companies are under constant cost pressure. One result: The supply chain functions of logistics, production, and supply management are managed as cost centers. You can thus expect your customers' purchasing professionals to be unrelenting in their efforts to reduce costs. Globalization has emboldened them by increasing factor mobility and market access, which gives them alternative sourcing options—options that often possess low-cost labor advantages.

To reduce costs, companies pursue a combination of four strategies:

- **Productivity enhancement**—The key to improving productivity is to promote learning.[9] Companies are thus minimizing work rules and increasing training (including cross-training). The goal is to empower workers to find and implement better ways to do things.

- **Automation**—Technology is making process redesign a perpetual initiative.[10] Many tasks that have been performed manually can now be automated. Others, especially those that can be digitized, can be relocated to lower-cost locations. New technologies such as additive manufacturing (also known as print manufacturing) will continue to enable process redesign.[11]

- **Global network design**—Improved logistics have reduced the total landed costs of products made in distant lands.[12] Companies are taking advantage of this by locating facilities in far-flung countries with low-cost inputs.[13]

- **Outsourcing**—Sometimes companies discover that they can no longer perform certain value-added activities as well as other members of the supply chain. When this happens, they move from make to buy. Outsourcing enables companies to do what they do best, relying on supply partners for complementary value creation.[14]

Carefully considering these four strategies reveals that a true cost capability cuts across value-added processes, requiring intra- and interorganizational collaboration. Processes across the entire supply chain from raw materials to end consumer must be designed for efficiency. One manager in the CPG industry emphasized that the real measure of cost performance is "total landed cost to the customer's trunk."[15] Effective cost reduction can initiate a powerful cycle of competitiveness. As they expand market share, companies can improve scale economies and raise profitability. The added funds can then be reinvested in future capabilities.

Quality

Quality is often defined as conformance to specs. A customer focus, however, means that quality is nothing more or less than meeting customer expectations. David Garvin, a Harvard professor, identified eight factors that customers use to assess quality:[16]

- **Performance**—A product's operating characteristics
- **Features**—The unique characteristics that distinguish a product from rivals' products
- **Reliability**—The user's ability to count on the product not to fail
- **Conformance**—How well a product conforms to design specifications
- **Durability**—A product's life expectancy (also, mean time between failure)
- **Serviceability**—The speed and ease of repair when problems occur
- **Aesthetics**—The perceptions of fit and finish (also, artistic value)
- **Perceived quality**—A product or brand's quality reputation

As for managing quality, Six Sigma (6σ) is the gold standard in quality philosophy and practice. Motorola launched Six Sigma in 1985. The goal was simple—eliminate defects. Motorola chose to call its quality program Six Sigma for two reasons. First, statistically speaking, a process that achieves 6σ quality produces only 3.4 defects per million parts made. Second, managers wanted to avoid the stigma attached to total quality management (TQM) programs. Six Sigma follows the DMAIC methodology—Define, Measure, Analyze, Improve, and Control—and applies statistical tools to identify and remove the causes of defects in value-added processes. As process variability is eliminated, quality improves. Jeffrey Immelts, CEO at General Electric (the company that made 6σ popular), calls Six Sigma the common language at GE. Everyone from the loading dock to the C-suite is expected to speak the language of Six Sigma. At GE, quality is important everywhere.

You may wonder where quality ranks among the five value dimensions in importance. For many decision makers, quality is more vital in deciding purchase decisions than cost. Quality has been called the most vital factor in long-term success. W. Edwards Deming went so far as to say, "You are not obliged to manage quality. You can also choose to go out of business."[17]

Delivery

Managers' penchant for lean, just-in-time operations has made a delivery capability an increasingly important source of differentiation. Companies frequently operate with minimal inventory—sometimes as little as two to four hours of supply. They rely on suppliers' ability to deliver to promise. They may even specify a delivery time window so that they can better plan their own operations and improve facility utilization. Thus, delivery is more than just "doing things fast." It is also the ability to do those things consistently. Fast, reliable delivery requires short order cycles and reduced variability.

If your company wants to use delivery as a competitive weapon, you must focus on building speed across a variety of processes. For example, in a classic example, Motorola became a world leader in pager manufacturing by reducing production cycles from 30 days to less than 30 minutes. National Semiconductor, by contrast, redesigned its global distribution network to reduce order fulfillment lead times by 47 percent. A 34 percent increase in sales resulted.[18] Customers value speed. Fast, reliable cycles improve forecast reliability and reduce the need to carry inventories. Sony de Mexico, for instance, reduced its overall order cycles by 75 percent, enabling its customers to cut their inventories by 50 percent.[19]

These examples illustrate the cross-functional nature of a delivery capability. Any activity that negatively impacts the time or variability of the order cycle diminishes a firm's ability to deliver on time. A late supplier delivery, a machine malfunction, a parts shortage, an incorrect order entry, or a transportation delay reduces delivery performance and drives costs up. For instance, an electronics manufacturer operating in the Dominican Republic experienced persistent production delays. To compensate, logistics managers resorted to airfreight to meet promised delivery schedules on 70 percent of orders—at a cost 600 percent higher than ocean shipping.[20] Research has shown that firms that experience supply chain delivery glitches report on average 6.92 percent lower sales growth, 10.66 percent higher growth in cost, and 13.88 percent higher growth in inventories.[21]

Responsiveness

Change is the only constant in today's business world. The ability to act quickly—that is, to adapt or respond—as customers make special requests, competitive requirements change, or the unexpected happens conveys a vital advantage.[22] Disasters such as the 2011 earthquake off the coast of Japan—and 2011 flooding in Thailand—have brought new attention to the need to build a responsive supply chain. Being responsive enables a company to deal with the risk inherent in lean, global operations. Responsive companies recover from the unexpected more quickly and resume operations faster than the competition. Sun Tzu summarized the power of responsiveness as follows, "Every minute ahead of the enemy is an advantage."[23]

Responsiveness, like the other value dimensions, is a cross-functional capability that relies on effective information systems, well-designed processes, and the adaptability of the firm's people.[24] The following steps are critical to creating a responsive culture:

- Make responsiveness a priority throughout the firm and across supply chain relationships.

- Map processes to make them visible and to identify responsiveness enabling activities or decisions. Use mapping to initiate risk-mitigation discussions and identify operating alternatives.

- Use information systems to monitor operations, link to customers, promote proactive environmental scanning, and share information on a real-time basis across the network.
- Cross-train workers and organize work in multifunctional teams.
- Design performance measures to value responsiveness.
- Build learning loops into every process throughout the organization.

Toyota has built a highly responsive supply chain. Suppliers are required to locate near Toyota's production facilities for fast, reliable, and flexible delivery. Within Toyota's factories, information and logistics systems synchronize materials flow to incoming orders—enabling custom assembly. Further, Toyota can build multiple models—for example, the Camry sedan, Sienna minivan, Highlander SUV, and Lexus RX 330—using the same platform and on the same production line, helping it mix and match production to market demand. If Toyota does not properly anticipate market needs, it can quickly adapt to them.

Innovation

To avoid "one-hit-wonder" status, companies need to be able to innovate consistently. In most industries, the time between new product introduction and product "obsolescence" has shrunk dramatically. Rivals can copy and introduce their own "new and improved" version of products within six to nine months, making first-mover advantage fleeting. Apple's iPad illustrates the innovation dilemma. Pushed by rivals like Amazon, Google, Samsung, and Sony, Apple has brought a new iPad to market every year since it introduced the original iPad in April 2010. The gap between the iPad 3 (March 2012) and the iPad 4 (November 2012) was a scant eight months. Research has long shown that bringing products quickly to market is the key to financial success.[25] In one instance, products introduced six months late, but within budget, realized a 33 percent decrease in expected profits over the first five years. Products introduced on time, but 50 percent over budget, realized only a 4 percent reduction in profit.[26]

Process innovation—often overshadowed by the quest for new products—can also deliver unique competitive strengths. Into the early 2000s, much of Dell's success resulted from patents, not for its products, but for various aspects of its manufacturing processes. By 2003, Dell had earned more than 550 business-method patents. Michael Dell's mantra was "Celebrate for a nanosecond, then move on." Industry analyst Erik Brynjolfsson noted, "They're inventing business processes. It's an asset that Dell has that its competitors don't."[27] Dell's eventual undoing was a failure to bring the right new products to market quickly. When asked about what surprised him about the computer industry, Erik answered, "I didn't see [tablets] coming."[28]

Walmart is another company that has emphasized "inventing" processes as a competitive lever. Persistent improvement of cross-docking methods over a 20-year time period ensures high levels of on-shelf availability at everyday low prices. Walmart continues to work with suppliers to improve the efficiency of its back-office operations (focusing on details like palletization and radio-frequency identification [RFID] tags) even as it strives to find ways to involve customers to reduce checkout times and improve the service experience. The key to process innovation is to cultivate a culture of experimentation and learning within the four walls of a company as well as across the supply chain. The benefits of these efforts include greater efficiency, enhanced quality, and faster cycles. One additional, overlooked benefit: Process innovation tends to be much more difficult to copy than product innovation.

Total Order Performance—A Synergistic Approach

As you read about the dimensions of customer value, you may have thought, "Based on my personal experience, not all value dimensions are equally important in every purchase decision." If so, you would be correct. Just as individual consumers weight each factor differently, companies likewise have distinct priorities. Even for the same entity, unique purchases are handled differently, depending on the importance or type of buy. Terry Hill, professor at London Business School, expressed these ideas using the language of order qualifiers and order winners.[29] To help you prioritize decisions regarding value creation, you will want to remember three rules:

- **Get into the game**—Across most purchase decisions, cost and quality are the critical value dimensions. If you want to be taken seriously as a potential supplier, you have to perform well in these areas. However, because they are universally important, high levels of parity often exist in these areas. Cost and quality thus tend to be order qualifiers.

- **Differentiate yourself**—Harvard's Michael Porter made a vital observation: Sustained success requires you to differentiate your company's value proposition in the mind of customers.[30] If your cost and quality positions are good enough to get you consideration as a supplier, you need to differentiate yourself along the lines of one of the other dimensions. That is, customers must view your delivery, responsiveness, and/or innovation as an order winner.

- **Avoid disqualification**—You must meet minimum requirements across all five value dimensions. Even if you rate well on cost, quality, and a differentiating characteristic, you could still disqualify yourself via unacceptable performance elsewhere. Your customers are keeping score. You earn points along each dimension. Having an order disqualifier will make it impossible to earn enough points to win (or keep) a customer's business.

You may also have noticed that efforts to create value in one area influence the other value dimensions. Many interactions exist within and among them. For instance, focusing strictly on quality, efforts to add cutting-edge features may hinder reliability. By contrast, efforts to design in serviceability may improve reliability and durability. Your challenge is to make the interactions more visible so that tradeoffs can be accurately assessed and good decisions made.

Importantly, until recently, managers believed it was impossible to concurrently pursue improvements across all value dimensions (see Figure 1-3). Managers viewed high quality as inherently expensive. They perceived consistent delivery to conflict with responsiveness. Leading companies have shown, however, that the relationships are more nuanced—and often synergistic. Better process visibility, information exchange, and workforce flexibility promote higher levels of performance across all value dimensions. For instance, the term *hidden plant* was coined to describe the fact that 15 percent to 40 percent of a firm's capacity is used to find and fix poor-quality work.[31] In many settings, better quality reduces costs. The ability to do things fast often can improve forecast accuracy and offer more decision gates. The result: Responsiveness improves and costs go down. A more holistic approach can enable the value dimensions to work together like the spokes of a wheel to advance the company to a more competitive market position.

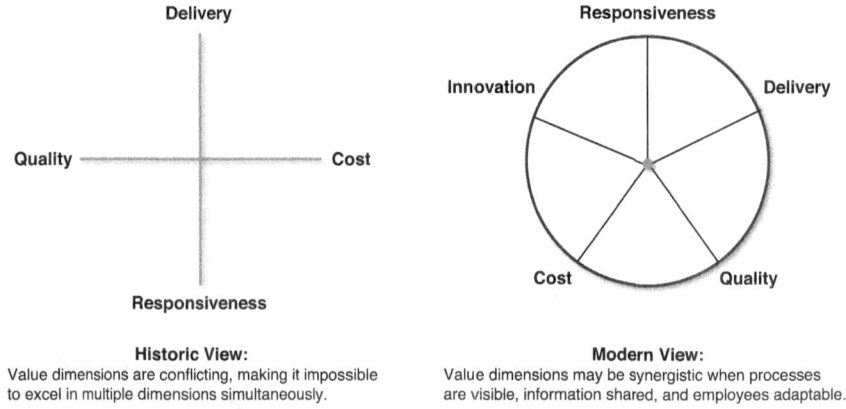

Figure 1-3   Interactions among value dimensions

Finally, you may have observed that logistics contributes differently to each value dimension. Table 1-1 summarizes logistics' value-creation role. Logistics is a core contributor to delivery and responsiveness capabilities. Indeed, logistics process and network design can enable a company to compete on speed and to mitigate the negative consequences of unexpected events. Such capabilities are hard for rivals to replicate. By contrast, logistics provides only tangential support to a quality advantage. However, poor logistics can become a quality disqualifier. As for costs, careful logistics management can make a direct contribution to a cost advantage, but other functions—especially purchasing—represent a much higher percentage of cost of goods sold (COGS). Thus, logistics can enable a cost advantage but not secure one. Finally, logistics innovation can help a company develop distinctive processes, improving delivery, responsiveness, cost, and quality performance. Logistics is a vital value creator.

Table 1-1   Logistics' Contribution to Value Dimensions

Value Dimension	Definition	Logistics' Contribution
Cost	The ability to achieve lower cost levels than rivals	Logistics costs often represent 8 percent of a company's COGS. By reducing total logistics costs, a company can lower total landed costs and be more competitive in expanded geographic markets.
Quality	The ability to deliver products and services that customers view as better than those offered by rivals	Logistics' main contribution is to avoid damaging product during shipment. However, unique or tailored services can improve customers' perceptions of quality service.
Delivery	The ability to deliver with consistently short cycles	Logistics is a core contributor to a delivery capability. Effective order cycle design and management reduces cycle times and ensures dependability. Inventory, transportation, and network design decisions can provide a delivery advantage.
Responsiveness	The ability to quickly respond to unexpected events	Logistics is a core contributor to a responsiveness capability. Process design augments inventory, transportation, and network design to help provide the adaptability and flexibility needed to respond to the unexpected.
Innovation	The ability to bring new products to market quickly or to improve processes consistently	Logistics seldom contributes meaningfully to product innovation. However, incremental and/or radical innovation in processes can provide distinctive advantage—typically in the areas of delivery and responsiveness.

Contributing to Customer Satisfaction

Customer satisfaction is the essence of exchange—that is, the reason customers buy things and build business relationships. To achieve high levels of satisfaction, your company must not just create outstanding value, but it must create *the* value customers desire and are willing to pay for. This is not a new idea. Rather, it is an idea that is recycled every few years. In fact, *Business Week's* March 12, 1990, cover was titled, "KING CUSTOMER." The tagline read, "Forget market share. Stop worrying about your competitors.

The companies that are succeeding now put their customers first." Putting customers first, of course, requires understanding how customers form perceptions of satisfaction.

Much of the customer satisfaction research has focused on expectancy disconfirmation.[32] In essence, people act the way they do because of what they expect to happen as a result of their choices.[33] Figure 1-4 illustrates the mechanics and implications of this idea. Satisfaction begins with each customer's a priori expectations. When she engages in a service experience, her expectations are either confirmed or disconfirmed. Three outcomes are common:

- Experiences that fail to meet expectations lead to dissatisfaction. If the experience is bad enough, the customer may complain (remember "United Breaks Guitars").
- Experiences that meet expectations deliver satisfaction.
- Experiences that exceed expectations achieve strong satisfaction—a result that may lead to loyalty and repeat business.

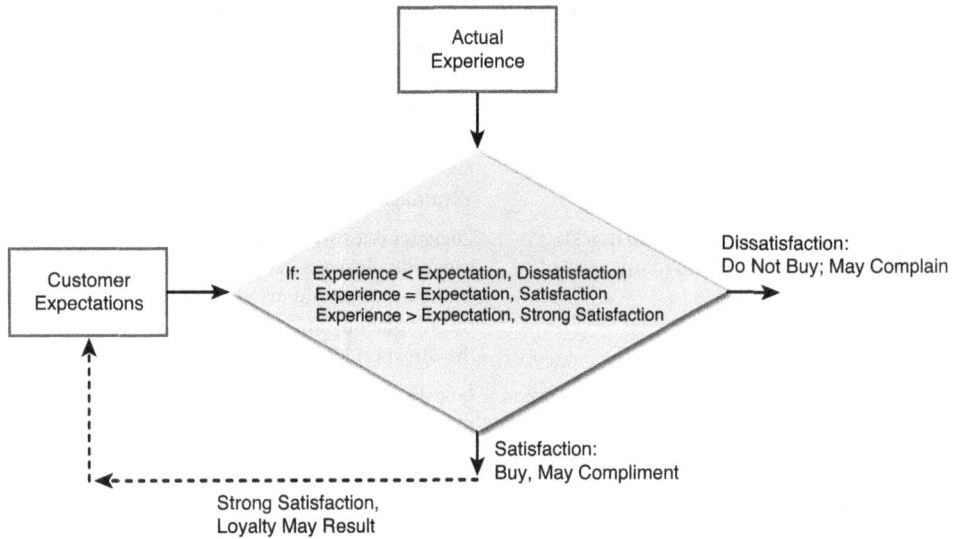

Figure 1-4 The formation of satisfaction perceptions

Let's dig a little more deeply into this process. Research has shown that expectations arise from a combination of cognitive and affective processes.[34] At the cognitive level, the human mind processes (i.e., captures, assimilates, and responds to) information related to each experience. Memories of past experience mingle with external stimuli (e.g., advertising) to help create a learned expectation.[35] The affective process is more emotional—it is a "gut" reaction. Because customer emotions arise quickly, they shape and are shaped by the cognitive process embedded in each experience.[36] The comingling of rational and emotional responses intensifies feelings of satisfaction/dissatisfaction and influences long-term behavior (e.g., purchase decisions). The bottom line: You must capture both the mind and the heart of your customers to achieve high levels of satisfaction.

Now that we understand the basics of satisfaction formation, let's briefly discuss the three core strategies companies employ to meet customer needs and achieve satisfaction.

Customer Service Strategies

At many organizations, the service focus is operational, focusing on things like product availability, order lead times, and reliability. The service goal is to perform to industry standards. Everyone in the industry—both suppliers and customers—knows what these benchmarks are. Standards are expressed as minimums (e.g., 90 percent on-time delivery) or as industry targets (e.g., 97 percent on-time delivery). Logistics executives commit to service levels that are equal to or slightly better than industry rivals. Most of the time, overall or average performance is tracked and used to assess performance and guide improvement initiatives. Because better service usually comes at a price, the top-of-mind question is, "Would better service really be worth the extra costs and provide a strong return on investment (ROI)?" Unless executives have consciously chosen to compete on delivery and responsiveness, an ever-present emphasis on costs will lead executives to commit to lowest service levels required to remain competitive.

Although common, basic service strategies possess several inherent flaws (see Table 1-2). Because they are operational and internally focused, managers must hope that by meeting industry standards, they are meeting customers' needs. Traditional service programs do not guarantee that managers truly understand customer needs—in aggregate or by segment/specific customer. Nor do they provide insight into whether customers are truly satisfied. Managers may feel they are delivering outstanding service—and they are, based on their internally generated metrics—when customers feel otherwise. Even industry-leading performance will not be rewarded if customers do not value the type of service being provided.[37] Misdirected service leads not only to wasted resources, but also to a reputation for mediocre service.

Table 1-2 Limitations of Different Customer Fulfillment Strategies

Strategy	Goal & Focus	Potential Downsides
Customer service	Meet/exceed industry standards Inward looking	Fail to understand what customers value. Fail to satisfy the customer. Expend resources in wrong areas. Measure performance inappropriately. Operational emphasis leads to service gaps.

Strategy	Goal & Focus	Potential Downsides
Customer satisfaction	Meet/exceed customer expectations	Ignore operating realities; overlook operating innovations.
	Outward looking	Giving customers what they say they want leads to product/service proliferation.
		Maintain unprofitable relationships.
		Historical focus makes the company vulnerable to disruptive products/processes.
Customer success	Help customers meet their customers' needs	Limited resources require that "customers of choice" be selected; that is, customer success is inherently a resource-intensive strategy.
	Forward/downstream looking	

Adapted from Fawcett, Ellram, and Ogden (2007)

To summarize, if they lose line of sight to customer expectations, basic service programs create service gaps. In one instance, managers at a European manufacturing facility set quality performance standards at industry levels, but lower than a key customer's expectations. Of note, the customer was a sister division within the same corporation. Shipments that met internal standards were returned as unacceptable. Frustration grew with repeated service failures. Better expectation alignment may have led to higher standards at the factory, increasing the short-term quality training costs, but would have reduced total relationship costs (inspection, inventory, return transportation, and rework costs would have gone down). Counterproductive intrafirm rivalry would also have evaporated. The lesson is clear: It is not enough for you to talk about the customer—you need to talk to the customer.

Customer Satisfaction Strategies

More companies today are talking to customers not only to learn about their needs, but also to communicate to them a firm commitment to fulfill those needs. They do more than talk about creating value; they promise to exceed expectations, delight customers, and fulfill dreams (see Table 1-3). Deep insight into customer needs is at the heart of satisfaction strategies. To gain this insight, companies conduct surveys, focus groups, in-depth personal interviews, and ethnographic studies. Figure 1-5 shows, however, that not all data-collection efforts provide equally valuable insights. Gathering the most insightful information is costly. Many companies thus rely on surveys and focus groups instead of much more expensive shadowing and ethnography. Yet, "living" with customers almost always provides more meaningful customer insight. Companies like Procter & Gamble install cameras in customer homes to watch how customers really use their products. In the B2B setting, senior executives are spending more time (up to 20 percent to 30 percent) with customers to gain an appreciation of customer needs.[38] Some companies even colocate their people on site at customers' facilities to really find out what customers are thinking, feeling, and striving to accomplish.

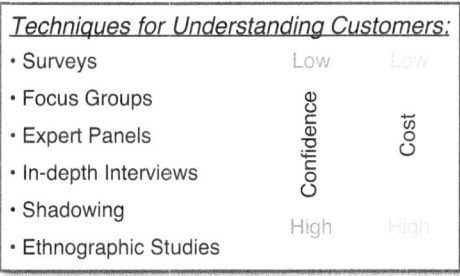

Figure 1-5   Comparison of cost and confidence for different data-gathering techniques

Table 1-3   Expressions of Customer Service, Satisfaction, and Success

Mission Statements That Express a Service Commitment

- At the heart of the corporate purpose, which guides us in our approach to doing business, is the drive to serve consumers in a unique and effective way.
- Our purpose is to create superior value for our customers.
- We have a vision: Customers For Life. To earn our customers' loyalty, we must listen to them, anticipate their needs, and act to create value in their eyes.
- Our customers are our partners in creating value; their loyalty is our greatest reward.
- We impress our customers so much that they want to buy again. We do it right the first time but "wow" our customers on recovery when we miss.

Mission Statements That Express a Satisfaction Commitment

- We exist only because our customers are satisfied and want to do business with us...and we never forget it!
- Customer satisfaction is the source of employee, shareholder, supplier, and community benefits.
- Our customers are the most important stakeholder in our business. Therefore, we go to extraordinary lengths to satisfy and delight our customers. We want to meet or exceed their expectations on every shopping trip. We know that by doing so we turn customers into advocates.
- We are committed to doing more than meeting our customers' needs. We strive to delight our customers by anticipating and exceeding their expectations through an innovative and creative workforce.
- We fulfill dreams!

Mission Statements That Express a Success Commitment

- We support our customers' success by creating exceptional value through innovative product and service solutions.
- Only by serving our customers well do we justify our existence as a business. We view our success as dependent on our customers' success, both now and in the future.
- We exist to solve problems for our customers.
- We will provide branded products and services of superior quality and value that improve the lives of the world's consumers.

Well-executed data-gathering initiatives answer the following questions:

- What value/experience do customers really expect? How do they define quality, delivery, responsiveness, and other key value areas?
- How do customers measure our performance? Are our measures consistent with theirs?
- How well does our performance meet our customers' expectations and requirements?
- In what ways could we improve performance to differentiate ourselves in customers' minds?
- Would customers really value better performance—enough to pay for it?[39]

Building on this insight, you can design your logistics service systems to deliver the experiences your customers expect and value, ensuring positive confirmation and high levels of satisfaction.

Now a warning: Customer satisfaction strategies possess limitations you need to consider. A misguided emphasis on satisfying customers can lead managers to make promises that cannot be fulfilled. When the service failure inevitably occurs, a promise has been broken and strong dissatisfaction emerges. Similarly, a desire to meet all customers' needs promotes excessive product/service proliferation, undermining operating efficiencies and performance. Too much stress on satisfying customers has led even world-class firms to unknowingly maintain and invest in unprofitable relationships. Another challenge arises from focusing too much on what has satisfied customers in the past. Rivals' disruptive product/service strategies can quickly persuade customers to switch allegiance. Of course, this assumes customers were loyal in the first place. Managers often mistake satisfaction for loyalty. In reality, satisfied customers may not be happy or loyal. Satisfaction means meeting expectations, not creating differentiation. To remedy this potential pitfall of satisfaction strategies, companies increasingly emphasize customer delight as their objective. Delight emerges as a company goes above and beyond customers' *existing* expectations (see Figure 1-6). Delight has been described as follows:

> Satisfaction is based on fulfilling the expected; delight is based on fulfilling the unexpected positive surprise-based occurrences. Satisfaction is based on meeting or slightly exceeding expectations, while delight occurs from features that are not expected or that add unexpected utility.[40]

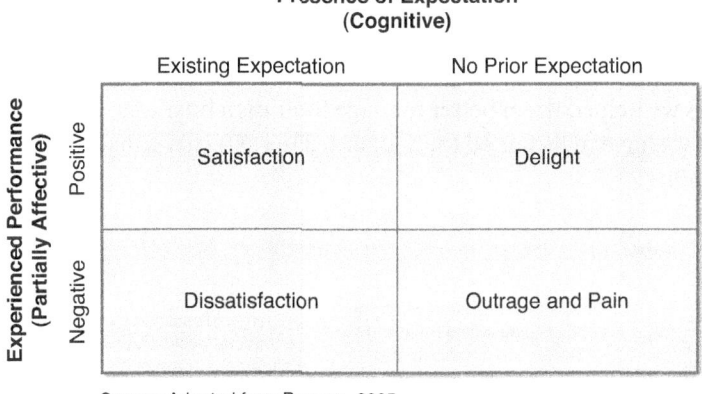

Source: Adapted from Berman, 2005

Figure 1-6 A framework for customer delight

Unexpectedly good experience—that is, happy, affirmative surprise—drives delight and is a precursor to differentiation and loyalty. To achieve delight, you must benchmark customers' typical experiences, understand their current expectations, and devise innovative, engaging service experiences. Please remember, however, that delighted customers are not always successful customers.

Customer Success Strategies

As Table 1-3 conveys, a small, but growing number of companies "view their success as dependent on their customers' success." General Electric's "At the Customer, For the Customer" and 3M's "6s at the Customer" programs exemplify customer success strategies. Managers at these companies realize the best way to drive growth and long-term competitiveness is to help their customers succeed. One CEO phrased the concept this way, "We turn our customers into winners. Their success is cash in our bank."[41] Well executed, a customer success strategy ensures loyalty even as it attracts new customers.

To be effective, managers must understand downstream supply chain dynamics—that is, what their customers' customers really want. Jack Kahl, CEO at Manco, emphasized this point, saying, "I have to know more about my customers than I know about myself."[42] Without this deep understanding of customer requirements, companies cannot provide the tailored service and mentoring that customers need to win their own competitive battles. Figure 1-7 reveals that the key to executing a customer success strategy is for managers to find ways to use their company's distinctive skills to help customers solve their own problems and improve their own capabilities. At PepsiCo, for example, a key customer was poised to switch to Coke for its fountain drinks. The customer saw no advantage to working with Pepsi. The logistics team saw things differently and sought an opportunity to change the competitive dynamics. Managers had noticed that the customer was struggling to manage routing on inbound shipments. A senior executive at Pepsi described what happened, saying, "We offered to share our experience in managing routing. As we helped them better manage their own business, their attitude changed positively and we have maintained the account. This external consultancy creates a sense of indispensability."[43]

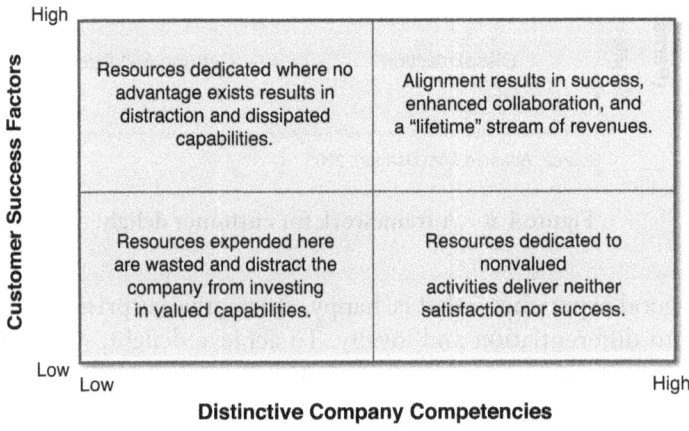

Figure 1-7   A Customer Success Framework[44]

As a game changer, customer success strategies are powerful, but they are not without real implementation challenges. For instance, the executive who led the Pepsi team described an assessment dilemma, noting, "We would have lost this account, but didn't thanks to our efforts to provide solutions. So, what do we document as gain? How do we justify the costs associated with helping our customers improve their own operations?" This comment alludes to the second challenge: Customer success strategies are resource intensive. Even the largest, most profitable companies do not have the unlimited resources to dedicate to being a consultant to every customer. You will need to apply success strategies selectively—to customers and situations that offer a real return on investment.

Service System Design

Regardless of your firm's choice of fulfillment strategy, success depends on how well you design and manage your customer-experience system to deliver a uniquely positive experience. Figure 1-8 introduces customer-experience design to the expectancy disconfirmation model. Everything still begins with a deep understanding of customers' initial expectations. Managers use this insight, along with a knowledge of their company's value-added capabilities, to develop attractive value propositions. Value propositions are the promises a company makes to customers about how it will meet their needs.[45] Value propositions serve two roles:

- They shape customer expectations, influencing customer purchase decisions as well as how the customer will assess the actual service experience.
- They define what the company must do to earn a customer's business, setting the parameters for the design of the company's customer-experience system.

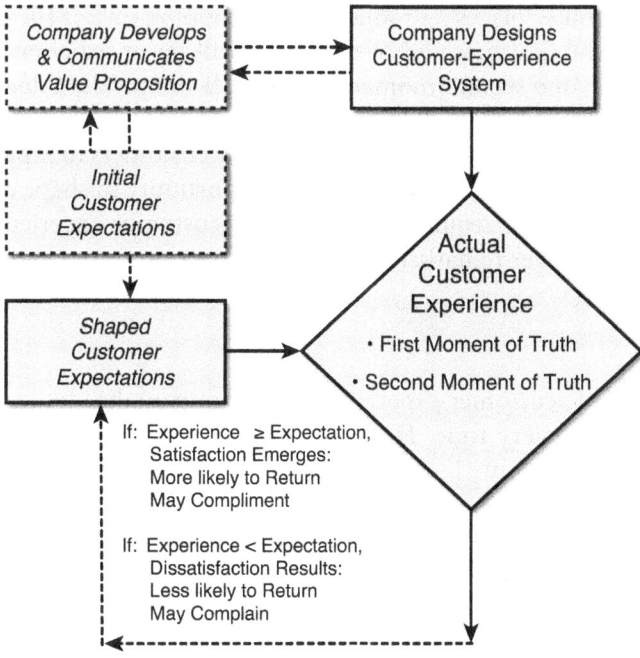

Figure 1-8   Service system design[46]

Touch Points

Satisfaction emerges as the customer touches a company's value system at various points of the experience. Jan Carlzon, CEO of Scandinavian Airlines, coined the phrase "moment of truth" to help employees grasp that each touch point is vital because it may be the moment that sets forever customer impressions of the firm.[47] Two types of touch points exist. Acquisition touch points, which are consistent with the economic concept of value-in-exchange, occur as customers learn about and make purchase decisions. For instance, as you make vacation plans, you likely compare alternative flight schedules and costs and then make flight reservations. Utilization touch points, the equivalent of value-in-use, occur as the service is experienced. In-flight service exemplifies a utilization touch point (as does lost luggage).

To design winning customer-experience systems, you must explicitly consider how each type of touch point influences customer assessments of satisfaction. A. G. Lafley, CEO at Procter & Gamble, emphasized that acquisition and utilization touch points are equally critical. He called them the "two moments of truth." The first moment culminates in the store, when the customer places a product in her shopping cart. If she does not choose a P&G product in the first moment of truth, she will never experience the product's performance in use—the second moment of truth. If the product does not live up to expectations, she will not buy the P&G product the next time she is in the store. Lafley recognizes that acquisition touch points—which allow customers to express their a priori value assessments—enable, or end, a company's opportunity to shape customers' satisfaction perceptions. Success requires that the entire customer-experience value system must be designed to deliver remarkable customer experience.

Orchestration

To deliver remarkable customer experience, your company has to do the right things right—the first time, every time. How you design your service system determines whether or not your company builds the right capabilities to earn—and keep—a customer's business. In all but the most basic satisfaction scenarios, value-creation capabilities are orchestrated capabilities. Simply stated, no company has sufficient resources and skills to do everything needed to meet customers' needs. Companies therefore rely on a network of suppliers, service providers, and customers for needed skills to fulfill the company's value proposition. Orchestration is the skill that enables companies to bring the resources of the network together. Orchestration consists of three core steps:

1. **Select team members**—Knowing what value you need to create, you need to identify the right players—those with key resources—to participate as members of your value-added team.

2. **Assign team roles**—Based on a correct understanding of each player's skills, you must assign the right roles and responsibilities to each team member to create optimal value.

3. **Build team cohesion**—You need to remember that having the right players does not mean they will play well together. You therefore need to invest in team chemistry by establishing the right relationships among team members.

Your firm's orchestration skills as well as the resources possessed by each member of the team will determine both how and how well you execute each of these three steps. It takes very different skills to execute orchestration's three steps. Selecting the right team members requires extensive scanning and comparative evaluation skills. These are largely analytical, left-brain skills. By contrast, building team cohesion requires careful coaching (and sometimes coddling) to get different members of the team to want and be able to play together. This requires strong right-brain collaboration and creativity.[48]

The third step is perhaps the most difficult. The nature and intensity of relationships varies greatly.[49] If, for example, a customer possesses deep insight into changing market demand, a collaborative planning relationship may be appropriate.[50] If a supplier possesses distinctive technical know-how, intense and early supplier involvement in product design may be appropriate.[51] This may include colocation of workers. Of course, some relationships are necessary but do not offer unique value-creation potential. These are better suited to fair and efficient arms-length transactions among members of the network. Ultimately, building the right team makes all the difference in creating a remarkable customer-experience system.

Value Gaps

Because of its centrality to service system design, let's reiterate a point you cannot afford to forget or misinterpret: Only the customer determines satisfaction. You may incorporate extensive customer feedback into the design of your company's customer-experience system and be confident you got everything just right, but the customer alone assesses the experience, providing the post hoc appraisal of value. At times, your company may deliver to promise, exceed industry standards, and meet customers' a priori expectations; however, having experienced both moments of truth, the customer determines that she overestimated the anticipated benefit. Disappointment and dissatisfaction follow—largely because an unexpected value gap emerged.

Because value gaps are so important, let's briefly discuss the different types of gaps that hinder service quality. Figure 1-9 identifies six gaps that affect satisfaction perceptions and can damage buyer/supplier relationships:[52]

- A **knowledge gap** often exists between customers' real expectations and your perceptions of those expectations. If you do not accurately assess what customers really want, you will fail to design the right customer-experience system. Gaps here perpetuate gaps throughout the design and execution of winning customer-experience systems. Investments elsewhere are destined to deliver less-than-desired results.

- Sometimes you possess accurate insight into customer needs, but fail to translate your understanding into operational standards. The **translation** or **specification gap** emerges when you focus too intently on industry standards or internal capabilities that are not aligned with customers' real needs. They are thus common in basic service strategies.

- Even if standards are set appropriately, poor execution can lead to a **performance** or **service delivery gap**—for example, the standard calls for 98 percent on-time delivery, but you only deliver 95 percent of shipments on time. Because service systems, especially orchestrated systems, are complex, you need to design for visibility and traceability. Constant effort is needed to identify and remove the root sources for service failure.

- When it comes to shaping expectations, some companies are their own worst enemies. Marketing may fail to communicate to logistics the promises it has made. More common, someone makes promises beyond the systems' capabilities. **Communication gaps** always emerge when you overpromise and underdeliver. These gaps sour a relationship quickly. Sadly, despite the fact that they can be easily avoided, they frequently occur.

- Although not in the original gap model, customers sometimes interpret operating results differently than reality. This **perception gap** can easily undermine the relationship. For example, customer measurement systems may not capture actual performance. Sometimes, customers place extra emphasis on the most recent service. You may feel intense pressure to *always* perform because your customers constantly tell you that, "You are only as good as your last performance."

- The final gap, known as the **satisfaction** or **service quality gap**, occurs when the customer's perception of service and expectations are not aligned. Because word of mouth, operating requirements, and past experience affect expectations, you must understand where customer expectations are coming from and work to proactively shape them. Constant measurement and active communication are key.

The key takeaway: Negative value gaps at any point in the customer experience can undermine repeat business. Service systems must be designed not only for outstanding performance, but also for reproducible performance.

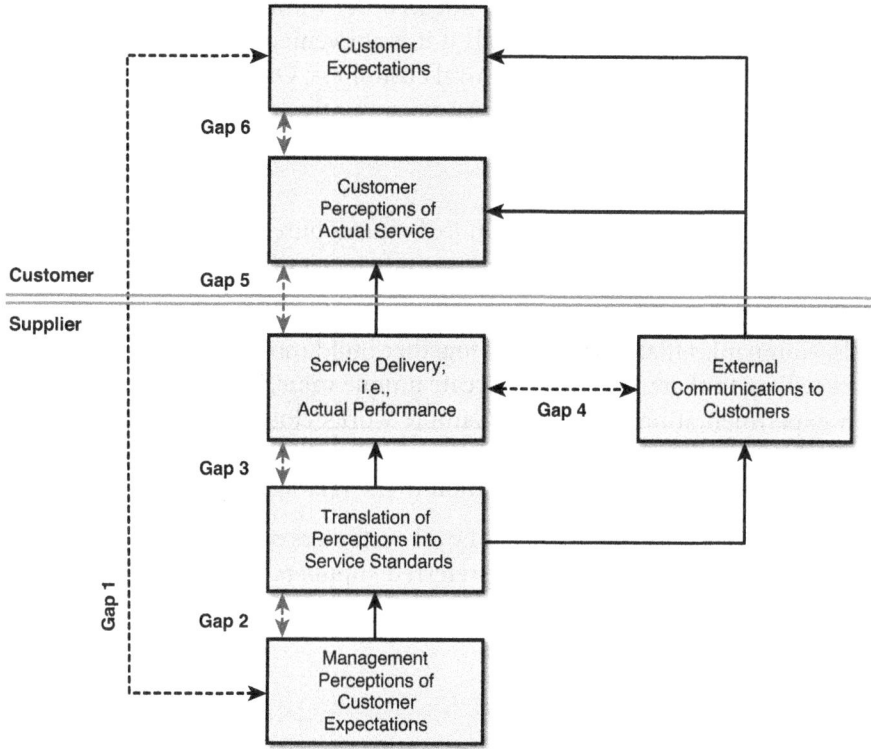

Figure 1-9 Understanding service value gaps[53]

Loyalty and Competitive Advantage

Your goal in designing your company's experience system is to achieve competitive advantage. Figure 1-10 ties customer experience to loyalty—a key precursor to advantage. Relationship marketing argues that it is often more important to retain customers—earning a larger share of their business—than to cultivate new customers.[54] One data point that supports this assertion is that it costs five to seven times more to acquire a new customer than it does to retain an existing one.[55] Cultivating loyal customers is, therefore, very important. Further, the experience attributes that drive loyalty also help drive new business.

Customer experience occurs along a continuum from inferior to comparable to distinctive. Customers defect when they experience inferior service. If companies don't change quickly, they go out of business in a hurry in our information-drenched economy. Comparable experience—that is, experience meets expectations and is viable vis-à-vis competitor offerings—delivers satisfaction. However, parity and indifference result. After all, such service acts as an order qualifier not an order winner. Although repeat business may occur, customers will defect to rivals if it is convenient. Too many managers mistake convenience-driven customers for loyal customers. Only when satisfaction springs from uniquely positive experience are customers motivated to establish closer, deeper relationships with a company.[56] These relationships are powerful for three interconnected reasons:

- They are characterized by larger, more frequent purchases—a fact that drives both top-line revenue growth and operating efficiencies (e.g., truckload shipping).
- They often lead to the design of innovative "tailored" services and business models. Companies that work closely together build more trusting relationships and are willing to share resources to create unique value. They often become partners in experimentation. Procter & Gamble works closely with both Walmart and Wegmans—two retailers that operate at vastly different scales—because they each bring process and product innovation to the relationship.
- They help recruit future business. Loyal customers often become guerilla marketers and evangelists—endorsing preferred suppliers and becoming part of these companies' marketing apparatus.[57]

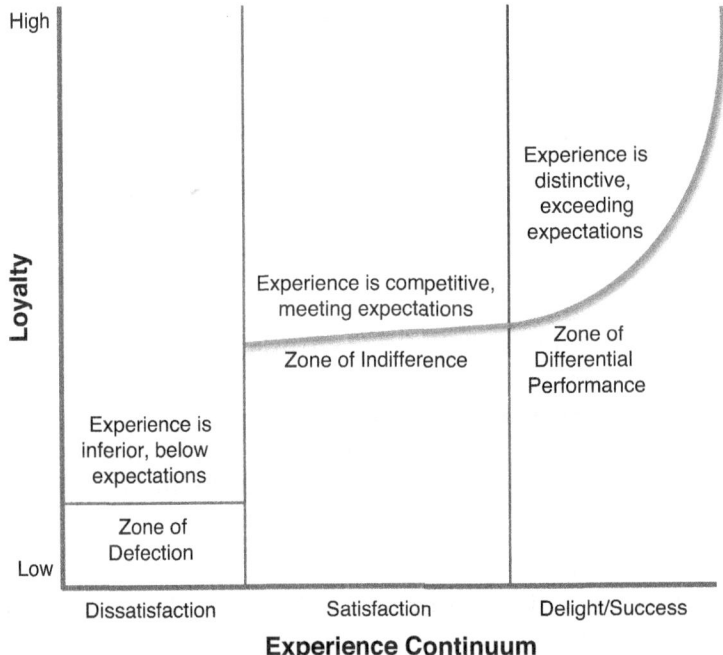

Figure 1-10 Customer experience, loyalty, and competitive advantage[58]

In summary, to profit from customer service, you need to provide customers a truly distinctive end-to-end customer experience. Both moments of truth must be remarkable. Building and managing such a customer-experience system creates loyalty. But, distinctiveness is very hard to achieve. Xerox, in a widely related study, asked customers to rate satisfaction on a 5-point scale. Largely satisfied customers—those who rated their Xerox experience a 4—were six times more likely to defect than those who rated Xerox a 5—completely satisfied/delighted.[59] The lesson: To achieve outstanding satisfaction ratings, you must build the systems that enable you to understand and influence customer needs—including those needs customers do not yet know they have. However, beyond meeting and exceeding customer needs, you must ensure that your value co-creation systems are both efficient and robust. Efficiency is needed to make sure that investments in service systems provide an adequate return. Higher service often, but not always, costs more. To charge more, you need to persuade customers that your service contributes to their own success in a meaningful way and is better than that of rivals. Robust means that you can deliver the positive experiences repeatedly over time. The inability to reproduce remarkable experiences will turn delight into disappointment as future experiences disconfirm elevated expectations.

Conclusion

As a supply chain manager, service system design is a core and vital aspect of your job. Your company's identity—and future—is defined by its ability to meet customers' real needs. Yet, despite its importance, few companies excel at creating remarkable customer experiences. Although found in the B2C setting, evidence of this challenge is seen in Figure 1-11, which shows almost 20 years of data from the American Customer Satisfaction Index (ACSI). The ACSI, which debuted with a score of 74.2 in 1994, remained essentially unchanged over the following 18 years. Given research shows that companies with high ACSI scores enjoy greater stock-price appreciation than their lower-scoring rivals, this finding is curious and disappointing.[60] The question arises, "Why is it so hard to improve customer satisfaction?" Two explanations persist:

- **Managerial commitment**—Remarkable experience requires intimate insight into customer expectations and supply chain requirements. It also requires unique and reproducible orchestrated capabilities. Despite the language in their mission statements, few companies have demonstrated sufficient commitment to execute both sides of the customer-experience equation.

- **Rising expectations**—Remarkable experience also hinges on a company's ability to deliver positive surprise. The customer experience must be different and better than the competition. If one company improves its value proposition (think Amazon.com's next-day free delivery), customers will adjust their expectations.

As rivals match the service offering, what once was remarkable becomes routine. To be consistently remarkable, you have to constantly improve.

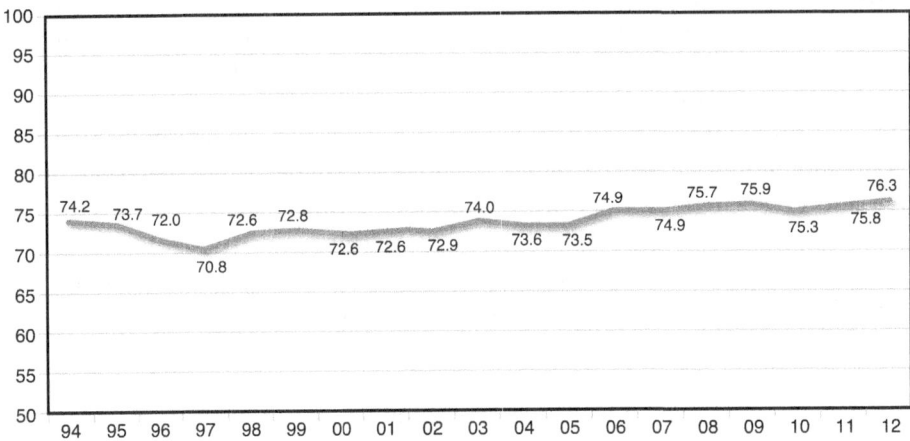

Figure 1-11   American Customer Satisfaction Index, 1994–2012 (4th quarter scores).

Actually, these two explanations intersect—it is very tough to remain committed and aptly invest in the strategy, structure, and systems needed to profitably offer remarkable service. As logistics order fulfillment is often the last touch a firm has with its customers, it can make or break your company's reputation, requiring you to understand fully the nature of order fulfillment.

Endnotes

1. Drucker, P. 2001. *The Essential Drucker*. New York: Harper Collins Publishers, Inc.

2. Henkoff, R. 1994. "Delivering the Goods." *Fortune* (November 28):64–78.

3. Fawcett, S., Jones, S., and Fawcett, A. 2012. "Breakthrough Trust: The Catalyst to Collaborative Innovation." *Business Horizons* 55(2):163–178.

4. Fawcett, S., Magnan, G., and Ogden, J. 2007. *Achieving World-Class Supply Chain Collaboration: Managing the Transformation*. Tempe, Arizona: Institute for Supply Management.

5. TheFind Web site, Retrieved on May 23, 2013, www.thefind.com/

6. See ThomasNet Web site, Retrieved on June 28, 2013, www.thomasnet.com/

7. Bowersox, D., Closs, D., and Cooper, M. B. 2012. *Supply Chain Logistics Management*. New York: McGraw-Hill/Irwin.

8. See, for example, Hayes, R., and Wheelwright, S. 1984. *Restoring Our Competitive Edge: Competing through Manufacturing*. New York: John Wiley and Sons; Hayes, R., Wheelwright, S., and Clark, K. 1988. *Dynamic Manufacturing: Creating the Learning Organization*. New York: The Free Press; and Hill, T. 2000. *Manufacturing Strategy: Text and Cases* (New York): McGraw-Hill Companies.

9. Skinner, W. 1986. "The Productivity Paradox." *Harvard Business Review* 64(4): 55–59.

10. Hammer, M. 1990. "Reengineering Work: Don't Automate, Obliterate." *Harvard Business Review* 68(4):104–131; Hammer, M. 2004. "Deep Change." *Harvard Business Review* 82(4):84–93.

11. "3d Printing: Out of the Box." *The Economist*, August 6. Retrieved August 21, 2013, from http://www.economist.com/blogs/schumpeter/2013/08/3d-printing; Tita, B. 2013. "How 3-D Printing Works." *The Wall Street Journal*, June 7. Retrieved September 16, 2013, from http://online.wsj.com/article/SB10001424127887323716304578483062211388072.html; Koten, J. 2013. "What's Hot in Manufacturing Technology." *The Wall Street Journal*, June 10. Retrieved September 16, 2013, from http://online.wsj.com/article/SB10001424127887323855804578510743894302344.html; Boulton, C. 2013. "Printing Out Barbies and Ford Cylinders." *The Wall Street Journal*, June 5. Retrieved September 16, 2013, from http://online.wsj.com/article/SB10001424127887323372504578469560282127852.html.

12. Fawcett, S., Ellram, L., and Ogden, J. 2007. *Supply Chain Management: From Vision to Implementation*. Upper Saddle River, NJ: Prentice Hall.

13. Scully, J., and Fawcett, S. 1994. "International Procurement Strategies: Opportunities and Challenges for the Small Firm." *Production and Inventory Management Journal* 35(2):39–46; Ferdows, K. 1997. "Making the Most of Foreign Factories." *Harvard Business Review* 75(2):73–88.

14. Fawcett, S., and Magnan, G. 2001. *Achieving World-Class Supply Chain Alignment: Benefits, Barriers, and Bridges*. Phoenix, AZ: National Association of Purchasing Management.

15. Ibid, 12

16. Garvin, D. 1983. "Quality on the Line." *Harvard Business Review* 61(4):65–75.

17. Deming, W. 1986. *Out of Crisis*. Cambridge, MA: MIT, Center for Advanced Engineering Study.

18. Ibid, 2

19. Ibid, 15

20. Ibid, 15

21. Hendricks, K., and Singhal, V. 2005. "Association between Supply Chain Glitches and Operating Performance." *Management Science* 51(5):695–711.

22. Bower, J., and Hout, T. 1988. "Fast-Cycle Capability for Competitive Power." *Harvard Business Review* 66(6):110–118; Bowersox, D., Calantone, R., Clinton, S., Closs, D., Cooper, M., Droge, C., Fawcett, S., Frankel, R., Frayer, D. Morash, E., Rinehart, L., and Schmitz, J. 1995. *World Class Logistics: The Challenge of Managing Continuous Change*. Oak Brook, IL: Council of Logistics Management; Stalk, G. 1988. "Time-the Next Source of Competitive Advantage." *Harvard Business Review* 66(4):41–51.

23. Tzu, S. 1963. *Art of War*. Oxford: Carendon Press.

24. Upton, D. 1995. "What Really Makes Factories Flexible." *Harvard Business Review* 73(4):74–84.

25. Wheelwright, S., and Clark, K. 1992. "Creating Project Plans to Focus Product Development." *Harvard Business Review* 70(3):70–82; Birou, L., and Fawcett, S. 1994. "Supplier Involvement in Integrated Product Development Strategies: A Comparison of U.S. and European Practices." *International Journal of Physical Distribution and Logistics Management* 24(5):4–14.

26. Clark, K., Chew, W., Fujimoto, T., Meyer, J., and Scherer, E. 1987. "Product Development in the World Auto Industry; Comments and Discussion." *Brookings Papers on Economic Activity*: 729–781; Clark, K. 1989. "Project Scope and Project Performance: The Effect of Parts Strategy and Supplier Involvement on Product Development." *Management Science* 35(10):1247–1263.

27. Park, A., and Burrows, P. 2003. "What You Don't Know About Dell." *Business Week* (November 3):74–84.

28. Worthen, B., and Anupreeta, D. 2013. "Inside Michael Dell's World." *The Wall Street Journal*, February 5.

29. Hill, T. 2000. *Manufacturing Strategy: Text and Cases* (New York): McGraw-Hill Companies.

30. Porter, M. 1980. *Competitive Strategy*. New York: The Free Press.

31. Feiganbaum, A. 1983. *Total Quality Control*. 3rd ed. New York: McGraw-Hill.

32. Oliver, R. 1980. "A Cognitive Model of the Antecedents and Consequences of Satisfaction Decisions." *Journal of Marketing Research* 17(November):460–469.

33. Oliver, R. 1974. "Expectancy Theory Predictions of Salesmen's Performance." *Journal of Marketing Research* 1974(August):243–253.

34. Oliver, R. 1997. *Satisfaction: A Behavioral Perspective on the Consumer*. New York: The McGraw-Hill Companies, Inc.

35. Tybout, A., Calder, B., and Sternthal, B. 1981. "Using Information Processing Theory to Design Marketing Strategies." *Journal of Marketing Research* 18:73–79; Edvardsson, B., Enquist, B., and Johnston, R. 2005. "Cocreating Customer Value through Hyperreality in the Prepurchase Service Experience." *Journal of Service Research* 8(2):149–161.

36. Lerner, J., and Keltner, D. 2000. "Beyond Valence: Toward a Model of Emotion-Specific Influences on Judgement and Choice." *Cognition and Emotion* 14(4):473–493.

37. Stock, J., and Lambert, D. 1992. "Becoming a World Class Company with Logistics Service Quality." *International Journal of Logistics Management* 3(1):73–80.

38. Fawcett, S., Magnan, G., and Ogden, J. 2007. *Achieving World-Class Supply Chain Collaboration: Managing the Transformation*. Tempe, Arizona: Institute for Supply Management.

39. Ibid, 38.

40. Berman, B. 2005. "How to Delight Your Customers." *California Management Review* 48(1):129–151.

41. Ginsburg, I., and Miller, N. 1992. "Value-Driven Management." *Business Horizons* (May–June):23–27.

42. Blackwell, R. 1997. *From Mind to Market: Reinventing the Retail Supply Chain*. New York: Harper Business.

43. Ibid, 38.

44. Fawcett, A., Fawcett, S., Cooper, M. B., and Daynes, K. 2014. "Moments of Angst: A Critical Incident Approach to Designing Value Systems for Outstanding Customer Experience." *Benchmarking: An International Journal* (in press).

45. Porter, M. 1985. *Competitive Advantage: Creating and Sustaining Superior Performance*. New York: Free Press.

46. Fawcett, S., and Waller, M. 2012. "Mitigating the Myopia of Dominant Logics: On Differential Performance and Strategic Supply Chain Research." *Journal of Business Logistics* 33(3):173–180.

47. Carlzon, J. 1989. *Moments of Truth*. New York, NY: Harper & Row, Publishers, Inc.

48. Fawcett, S., Andraski, J., Fawcett, A., and Magnan, G. 2010. "The Indispensible Supply Chain Leader." *Supply Chain Management Review* 14(5):22–29.

49. Richey, R., Roath, A., Whipple, J., and Fawcett, S. 2010. "Exploring Governance Theory of Supply Chain Integration: Barriers and Facilitators to Integration." *Journal of Business Logistics* 31(1):237–256.

50. Smith, L., Andraski, J., and Fawcett, S. 2011. "Integrated Business Planning: A Roadmap to Linking S&OP and CPFR." *Journal of Business Forecasting* 29(4): 4–13.

51. Petersen, K., Handfield, R., & Ragatz, G. 2005. "Supplier Integration into New Product Development: Coordinating Product, Process and Supply Chain Design." *Journal of Operations Management* 23(3–4):371–388; Froehle, C., and Roth, A. 2005. "New Measurement Scales for Evaluating Perceptions of the Technology-Mediated Customer Service Experience." *Journal of Operations Management* 22:1–21.

52. Parasuraman, A., Zeithaml, V., and Berry, L. 1985. "A Conceptual Model of Service Quality and Its Implications for Future Research." *Journal of Marketing* 49(4):41–50.

53. Ibid, 51

54. Peppers, D., and Rogers, M. 2006. "Return on Customer: A New Metric of Value Creation." *Journal of Direct Data, and Digital Marketing Practice* 7:318–321.

55. Hart, C., Heskett, J., and Sasser, W. 1990. "The Profitable Art of Service Recovery." *Harvard Business Review* 68(4):148–156.

56. Innis, D., and Londe, B. 1994. "Customer Service: The Key to Customer Satisfaction, Customer Loyalty, and Market Share." *Journal of Business Logistics* 15(1): 1–27; Rust, R., and Oliver, R. 2000. "Should We Delight the Customer." *Journal of the Academy of Marketing Science* 28(1):86–94.

57. Godin, S. 2003. *Purple Cow*. New York, NY: Penguin Group; McMullan, R., and Gilmore, A. 2008. "Customer Loyalty: An Empirical Study." *European Journal of Marketing* 42(9/10):1084–1094.

58. Ipsos. 2005. Loyalty Myth #8. *Loyalty Myths*, Retrieved September 16, 2013, from www.ipsosloyalty.com.

59. Motroni, H. 1993. "Company Study: Putting the Customer First." *Journal of Business and Industrial Marketing* 7(4):29–32; Fierman, J. 1995. "Americans Can't Get No Satisfaction." *Fortune* (December 11):186–194.

60. Fornell, C., Johnson, M., Anderson, E., Cha, J., and Bryant, B. 1996. "The American Customer Satisfaction Index: Nature, Purpose, and Findings." *Journal of Marketing* 60(October):7–18.

2

FULFILLING ORDERS: THE NATURE OF MODERN ORDER CYCLE MANAGEMENT

Opening Story: Speed Can Make Money

October 17

As David settled in for the long flight home, he pulled out his laptop. It had been three weeks since Diane had tasked him with reviewing and reimagining DWC's customer fulfillment capabilities. She had emailed him the previous day, asking him to put together a quick update by the end of the week. David sighed deeply—his "day" job didn't leave much time for setting up and running a new task force. He began to review the steps he had undertaken so far. His notes were organized using three questions to guide the reimagination efforts:

Where are we?

- Documented current performance vis-à-vis standards. DWC was hitting its targets.
- Benchmarked current performance against rivals. DWC was doing well. He had not been guilty of false advertising when he told Diane that DWC was an industry leader.

Where do we want to be?

- Identified order fulfillment benchmarks:
  1. The Supply Chain Council's SCOR model documented the standard order cycle.
  2. Anecdotal cases showed how other companies had addressed fulfillment crises.
- Put together a list of key customers to meet with. He wanted to know what they expected from DWC, how they perceived DWC's current performance, and how they measured DWC.

How are we going to get there?

- Put together a four-person team to work on the task force. The team included Paul Osterhaus from IT; Trina Cody, a direct report to Doug Hassle and lead for the account management team responsible for the Monster relationship; and Lisé Johnson, a financial analyst who had worked on a variety of supply chain projects in the past.

As tired as he was, David's gaze settled on the anecdotes. One in particular caught his attention: Sony de Mexico. On his third call to investigate order cycle best practices, a colleague had joked, "Just be grateful you're not Sony de Mexico. Poor order fulfillment raised their costs and just about shut them down. Only SCM^2 saved them." That statement had caught his attention. David had never heard the term SCM^2 before. He had asked his friend to tell him more. David reviewed his notes from the discussion:

- Sony's corporate headquarters had decided to shift capacity to Asia where labor costs were a fraction of those in Mexico. Low labor rates, minimal tariffs, and geographic proximity had not been strong enough reasons to justify Sony de Mexico's continued existence. *(Note to self: What do we do at DWC to justify our existence?)*

- While others scoffed at the idea that Sony de Mexico could survive, Rey, Sony de Mexico's CEO, had made a final attempt to save the operation. Out of desperation, Rey and his team had adopted the Six Sigma mantra of, "Forget what you think you know and let the data prove it to you." He reasoned that a blank-slate approach was the only way to change the destiny of Sony de Mexico.

- The breakthrough insight came from the voice of the customer. What could Sony de Mexico offer that Sony in Asia couldn't? When asked, customers repeated two facts:

- Based on existing performance levels—not much! Despite the close location—just across the U.S./Mexico border—Sony de Mexico's order fulfillment cycle was eight weeks. Asian operations could meet or beat that! Dealers were clearly baffled and frustrated.

- Dealer costs were high! Because they couldn't anticipate consumer demand across Sony's broad product line, dealers carried huge, expensive inventories.

(Note to self: Talk to DWC's customers. Find out what they value. What are their pain points?)

As Rey had processed the customer complaints, he had realized that his team also griped frequently about poor dealer forecasts. *(Note to self: We do the same thing.)* As a result, Sony de Mexico carried a lot of inventory—60 days of sales. The common denominator driving frustration for both Sony and its customers was long order delivery cycles. Rey had adopted the mantra: "Instead of complaining about forecast accuracy, let's build a supply chain robust enough to meet customer needs despite poor forecasts." *(Note to self: That's our challenge.)* Rey's team had to find out why lead times were eight weeks.

Ultimately, Sony de Mexico had reduced lead times to two weeks—a 75 percent improvement. Both dealers and Sony de Mexico were able to slash inventories. Most important, Asian operations couldn't match the shorter lead times that enabled vastly improved customer performance. This fact—that **S**peed **C**ould **M**ake **M**oney (SCM^2)—had saved Sony de Mexico.

Almost imperceptibly, David realized that the insight he was looking for was contained in that acronym: SCM^2. He now knew the next step and quickly typed, "Map out our actual order cycle. What drives our lead time? What drives the variability that causes us to drop the ball?" With that epiphany, David tucked away his laptop and leaned his seat back. He wouldn't arrive home until almost 10:30 p.m. A short nap wouldn't hurt.

Consider as you read:

1. What do you think of David's approach of evaluating the "as-is" and "to-be" states of DWC's order cycle? What performance should a well-designed order cycle deliver?

2. What are the key activities/steps that must be managed to achieve world-class order fulfillment? What causes order cycles (and other processes) to become inefficient or unreliable?

3. What are the tradeoffs David and his team can expect as they reimagine DWC's order fulfillment capabilities?

Fulfilling Orders

> "In retailing, the biggest single customer-service complaint is not having the item. If Kohl's is promoting Dockers at 25% off this week, you'd better believe the pants will be in stock. Otherwise, it's like inviting someone into your house and not offering him a seat."
>
> —Kohl's Executive

Have you ever wanted to buy a product, seen it advertised in a retailer's circular (in print or online), and rushed to the store to buy it—only to find the shelf empty? After you looked on the shelf, did you check the aisles, and maybe even ask a store employee for help? Ultimately, you were told the product was out of stock. At that moment, how did you feel? Were you delighted or, more likely, disgruntled? You are not alone in your sense of dismay or in your experience. Advertisements are an implicit promise that the store will have the product available for purchase at the promoted price. In fact, the Federal Trade Commission (FTC) issued a rule in 1971 stating that such a stockout occurrence is an "unfair or deceptive" practice. The ruling stated:

> [It is an] unfair or deceptive act or practice for retail food stores to (1) offer products for sale at a stated price, by means of any advertisement disseminated in the area served by any of its stores which are covered by the advertisement, when they do not have such products in stock and readily available to consumers during the effective period of the advertisement; and (2) fail to make the advertised items conspicuously and readily available for sale at or below the advertised price.[1]

Despite the implied promise, research has shown that retailers fail to have advertised product on the shelf 16.5 percent of the time (that is a one in six chance you will be disappointed). A more lenient, in-store measure (perhaps the product is on a special display or in the backroom) found advertised product out of stock 12 percent of the time.[2] This same study found nonadvertised items out of stock 7.6 percent of the time—a number consistent with other studies that report out-of-stock rates from 7 percent to 10 percent.[3] In case you are wondering, catalog retailers performed at a lower level with out of stocks occurring 15.9 percent of the time.[4] Calculating comparable numbers for online retailers is difficult. They do not have to actually possess a product to promise delivery. When they do not have inventory in the local DC, many online retailers simply don't tell customers that the product is not available. Rather, they transship from a distant warehouse or drop-ship from a supplier. However, if the online retailer wants to offer convenient, cost-effective delivery, low "on-shelf" availability undermines competitiveness.

Now, the vital question, "Why should you care?" First, if you think back to our discussion of the two moments of truth from Chapter 1, "Meeting Customers' Real Needs:

The Nature of Service System Design," you know that if a company does not win the first moment of truth, it will never have the opportunity to win the second moment of truth. The failure to have the product on the shelf invites the customer to switch brands or to shop elsewhere. In an online setting, it only takes a few clicks to find an alternative retailer. Second, out of stocks are expensive. It has been estimated that out-of-stock occurrences reduce overall sales by 4 percent, costing worldwide retailers about $435 billion in 2010.[5] Third, out-of-stock occurrences hurt the brand. Product availability—the result of a firm's order fulfillment capabilities—is the last touch point that customers have with their chosen suppliers. A failure here—that is, a late delivery in the B2B setting or an empty shelf in the B2C context—can persuade customers to find another supplier. We will take a closer look at out-of-stock costs later in this chapter.

To summarize, time and place failures are expensive order losers. What does this mean for you? You need to be able to design and execute an order fulfillment strategy that meets customers' needs efficiently and profitably. To do this, you need to answer three core questions:

- What do we want to achieve through our order fulfillment capability?
- How do we actually do this? (That is, what does a winning order fulfillment system look like?)
- What tradeoffs must we deal with as we develop our order fulfillment capability?

The Deliverables of an Order Fulfillment System

To design a winning order fulfillment capability, you need to begin with the end in mind. What do you need your order fulfillment system to accomplish? To define this, ask customers what they expect. Collect and scrutinize their supplier scorecards. Look downstream to your customers' customers and ask, "How can we help our customers perform better so that they win more business?" Generally speaking, customers require some mix of the traditional seven rights of logistics: the *right* product in the *right* condition and *right* quantity at the *right* time and *right* place for the *right* price to the *right* customer. If we translate these requirements into operational targets, your order fulfillment system will need to provide the following:

- Product availability
- Timely—i.e., fast, consistent, flexible—delivery
- Transparent, reliable service
- Service recovery
- Efficient operations

Product Availability

Availability is the most basic output of an order fulfillment system. Do you have the product available to deliver to customers when and where they want it? As discussed previously, companies often go to great pains to create demand and then fail to have the product available to sell. Nobody wins when this happens. Careful planning can help you avoid this situation.

Availability depends on many upstream activities such as supplier capacity, service-provider delivery performance, and manufacturing schedules. However, your company's inventory policy has the greatest influence on day-to-day availability. How much inventory is your company willing to hold in order to fill customer orders? Because inventory costs money, you cannot simply say, "We will meet 100 percent of all orders." This policy would be prohibitively expensive. To arrive at a reasonable inventory stocking policy, you need to consider the tradeoffs between inventory and stockout costs (this analysis is discussed later in this chapter). Tradeoff analysis begins with recognizing that not all products should be treated the same. Some products are critical—a customer might buy a car missing a cup holder, but not one without brakes. The rule for critical items is never to run out of inventory. Other products, although not critical, might be highly profitable but also easily substituted. A stockout means lost sales and profits; therefore, you should carry extra inventory.

To reduce overall inventory without sacrificing availability, many companies have turned to more sophisticated, often collaborative, planning and forecasting approaches. As part of its Collaborative Planning for Forecasting and Replenishment (CPFR) program, Best Buy employs weekly cadence calls with suppliers to discuss forecasts and promotional plans. The goal is to help everyone work off a single forecast that is developed using the best available data. Walmart provides suppliers point of sales (POS) information so they know exactly what is selling. Data analytics, often called "Big Data," is being used to sift through huge volumes of data to gain insight into what drives demand. For instance, customer buying habits, obtained through loyalty cards, are combined with external data on such things as weather or demographics to identify which products sell under different circumstances. By leveraging closer working relationships, more open information sharing, and better analytics, product availability is improving.

Companies assess availability by looking at stockout frequency, fill rate, and orders shipped complete:

- **Stockout frequency**—A stockout occurs when product is not available to fill a customer order. Stockout frequency indicates the percent of buying incidences in which stockouts occur. As related earlier, products at the retail level are out of stock 7 percent to 10 percent of the time (16 percent for advertised products). Remember, however, that a real stockout does not occur until a customer tries to buy a product. You should remember one additional caveat: Aggregate stockout frequencies do not consider product criticality or customer importance!

- **Fill rate**—A fill rate is defined as the percent of orders filled from stock on hand. Fill rates can be measured at the item, line, and order level. Importantly, a company can ship the same items, but report very different fill-rate percentages. You need to understand how distinct fill-rate metrics are calculated and what the numbers mean. Chapter 6, "Assessing Performance for Success and Improvement," discusses this in detail and presents and exemplifies performance metrics. Like stockout frequency, fill rates do not account for product criticality or customer importance.

- **Orders shipped complete**—As the name implies, orders shipped complete is a measure of your ability to ship complete orders to customers. If a single item is missing, the order is not complete and is recorded as a zero for calculation purposes. In reality, this is what the customer expects. Complete orders enable smooth operations, are easy to track, and do not generate extra, unnecessary costs.

Timely Delivery

To plan operations—for production and/or sales—customers depend on your ability to deliver their orders on time. The customer goal is to minimize costs, maintain efficient operations, and support sales. To achieve these goals, many companies now specify delivery time windows—some as short as 15–30 minutes. Toyota, for example, only carries 2 to 4 hours of inventory to support production. To ensure that its lines run smoothly, Toyota expects suppliers to deliver multiple times per shift. Toyota also asks many suppliers to sequence product in delivery racks so that the racks arrive just in time, are moved straight to the assembly line, and the parts match up with the vehicle being produced (Just In Sequence delivery). To make synchronized delivery possible, Toyota shares exact production schedules with suppliers. However, shared information is not sufficient to protect against production disruptions. To ensure consistency, Toyota has asked suppliers to locate their facilities within 200 miles of Toyota's assembly operations, to reduce production lead times, to ship via dependable carriers, and if necessary to hold extra inventory. Given a line stoppage can cost an automaker $10,000 to $100,000 per minute, neither Toyota nor its suppliers can afford to shut a line down because of a delivery failure.[6] For companies like Walmart that operate cross-dock DCs that rely on precisely synchronized inbound and outbound flows, on-time delivery is just as important.[7]

The Toyota example highlights two points regarding an on-time delivery capability. First, timely delivery is a cross-functional capability—it is achieved only as everyone involved in the upstream supply chain performs to promise. If a supplier's own production experiences delays, on-time delivery is jeopardized. This is true for packaging, warehousing, transportation, and anybody else who touches the product in fabrication or en route. In a global setting—where freight forwarders, global carriers, customs brokers, and other third-party logistics (3PLs) providers are involved—the number of touches and handoffs increases. A glitch at any point can create a costly disruption. Second, customers care about three aspects of delivery: speed, consistency, and flexibility:

- **Speed**—Speed refers to the length of the order cycle; that is, how long it takes from order placement to order receipt. Fast cycles reduce a customer's need to hold inventory. They also give customers more time to read the market, improving forecast accuracy to match supply to emerging demand. However, faster delivery can cost more. You may have experienced this when you requested overnight delivery from Amazon.com. Indeed, after seeing the shipping rate, you may have decided five-day delivery was good enough. You need to carefully assess customers' real needs. Offering faster delivery than customers need and are willing to pay for is the dark side of speed.[8]

- **Consistency**—Consistency is synonymous with dependability and may be more important than speed. If you can count on a supplier to deliver on time, you can plan with confidence. Known lead times mean no extra inventory is needed to compensate for the uncertainty caused by variable order cycles. In delivery, like in quality, variability is a costly enemy. Companies evaluate consistency in two ways. The classic approach is to compare actual order cycle times with planned order cycle times. Increasingly, however, customers care more about on-time delivery. Did you deliver when you said you would? For a customer like Toyota, this is what matters!

- **Flexibility**—Flexibility refers to your company's ability to effectively respond to the unexpected. For example, natural disasters may disrupt traditional supply lines. Can you still deliver on time? Or, a customer may request expedited delivery to support a surprisingly successful promotion. Can you deliver product quickly enough to spur the momentum? Events that require a flexible delivery capability may originate at the customer, in the upstream supply network, or in nature as follows:

    - Modification to the service agreement (e.g., change in ship-to location)
    - Sudden surge or decline in customer demand pattern
    - Support for unique promotional programs
    - New-product introduction

- Supply disruption (e.g., earthquake, storm, supplier bankruptcy, materials shortage)
- Product recall[9]

Transparent, Reliable Service

One thing to keep in mind as you design your order fulfillment system is that nobody—especially customers—likes a disappointing surprise. Ensuring availability and being fast, dependable, and responsive is of little value if picking errors mean the wrong product is shipped or if the correct product is shipped but arrives unfit for use. Customers rely on suppliers to perform a variety of logistics tasks hassle free. The perfect order concept emerged to capture and communicate this expectation. A variety of "perfect order" definitions exist; however, a comprehensive conceptualization is that a "perfect" order is received, processed, picked, packed, shipped, documented, and delivered on time without damage. At a minimum, companies define a perfect order as on time, complete, damage free, and correctly documented. An error in any of these areas creates an imperfect order. Reliable, worry-free service means that once a customer places an order, she does not need to be contacted again until it is delivered—and then only for routine receipt and use.

Providing customers real-time awareness of their order status enhances the perception of reliability. Today's technology, including bar codes, RFID tags, and satellite tracking, is making such transparency a reality. You've seen the technology in action when you've purchased a product from Amazon.com. With a few clicks, you can see when your order shipped, find out when it will be delivered, and track its progress each step of its journey to your doorstep. Customers value this transparency. When managers know everything is in order, they can manage with confidence. By contrast, when something goes wrong, managers routinely say that if they know that something is amiss with an order, they can proactively make adjustments.[10] Information shared early can help the customer reduce the costs of a service glitch. Of course, good—that is, relevant, timely, and accurate—information is helpful beyond warning customers of an impending glitch. Modern information technologies enable efficient, transparent supply chain operations and strengthen supply chain relationships. Good information improves planning, execution, and evaluation.

Service Recovery

In real life, no company achieves 100 percent perfect orders. Recognizing this fact, you can take steps to recover when, not if, a service failure occurs. In some instances, you may be able to find and execute a solution without the customer ever knowing that your order fulfillment systems experienced a glitch. For example, if a stockout occurs at the local DC, you may be able to transship the item from a more distant DC using expedited transportation and still deliver on time. When such a behind-the-scenes solution is not available, you should inform your customer immediately that a problem exists and offer alternative solutions. Being up front earns credibility and the two of you can work out an appropriate, cost-effective resolution.

You may be interested to know that research has shown that companies that aggressively resolve problems can avoid a negative hit to their service reputations.[11] In fact, the "service recovery paradox" exists because some companies have so successfully resolved service failures that customer satisfaction actually increased following the problem. Customers note that empathy and proactivity are critical positive influencers in the resolution process. It also helps if customers perceive the source of the problem to be out of your control.[12]

To reduce the incidence and mitigate the negative effect of service breakdowns, you should employ a contingency planning approach.[13] A well-conceived plan will help you anticipate potential problems—keeping some from ever happening—and plan for how you will resolve those that do occur. With a plan in place before a service failure occurs, you will be able to respond more quickly and effectively, improving overall perceptions of your service levels and helping maintain positive customer relationships. As Figure 2-1 shows, a typical contingency planning lifecycle consists of six core steps:

1. **Scan operating environment**—Identify potential opportunities for service failures.

2. **Risk assessment**—Evaluate potential consequences of specific service failure. Ask three questions: (a) How likely is each service failure? (b) What are its potential impacts? and (c) How prepared is our order fulfillment system to deal with the failure?

3. **Recovery plan**—Develop a service recovery plan for the service failures that are most likely to happen or that will have the greatest cost impact (be sure to include relationship costs in the analysis). For example, which is the best alternative DC for transshipping? What are the costs of different expedited transportation options? Importantly, you need to train and empower employees so they can act quickly and proactively to most standard or expected service failures.

4. **Enact/execute**—Implement the recovery plan when a service failure occurs.

5. **Evaluate**—Develop an after action or corrective action report that assesses the effectiveness of the recovery plan. Be sure to include a cost benefit analysis.

6. **Improve**—Update the recovery plan based on the after action plan and the customer's response and feedback. Given today's dynamic marketplace, contingency planning for service recovery should be done on a periodic basis.

Figure 2-1   Contingency planning process

Efficient Operations

Many companies pursue a quest to constantly improve service levels. Although better service is the theoretical ideal, you need to recognize that it does not always make sense to improve service levels. Strategically speaking, even if it were doable, you might not want to achieve 100 percent perfect orders. Higher service tends to cost more. Airfreight, for instance, costs more than shipping by containership—or even by rail or motor carrier. Locating a warehouse next to each customer is seldom economically feasible. The reality is that cost-service tradeoffs exist throughout the order fulfillment process. Your mandate is to efficiently provide the service customers need. Vitally, you need to remember that different customers want, and are willing to pay for, different kinds and levels of service. They, however, always expect to pay as little as possible for the services they seek. You, therefore, need to assess the cost and benefits of each service offering, seeking to take costs out of order fulfillment when and wherever you can.

The Details of an Order Fulfillment System

To establish a winning fulfillment capability, you need to understand the elements and mechanics of the typical order cycle (see Figure 2-2). At the broadest level, the order cycle is the elapsed time from recognition of need until the product is delivered and available for use. By definition, the order cycle begins and ends with the customer. The customer recognizes a need, prepares an order, and transmits it to the supplier. As a supplier, you receive, process, prepare, and ship the order. The cycle concludes when the customer receives, accepts, and pays for the product. Three touch points bring customer and supplier "systems" into contact, offering trading partners opportunities to work together to improve order cycle performance:

- **Collaborative planning, forecasting, and replenishment** can be used to synchronize expectations and operations via periodic cadence calls, which are used to share forecasts, discuss promotions, and communicate potential capacity imbalances and other disruptions that might adversely impact order fulfillment.
- **Order placement** consists of order transmittal and order receipt. Although this can be done manually via phone, email, or even fax, integrated electronic data interchange systems are frequently used to minimize costs and errors.
- **Order delivery** occurs when a shipment arrives at its destination, where the customer inspects and receives it. The customer takes ownership of and pays for the order.

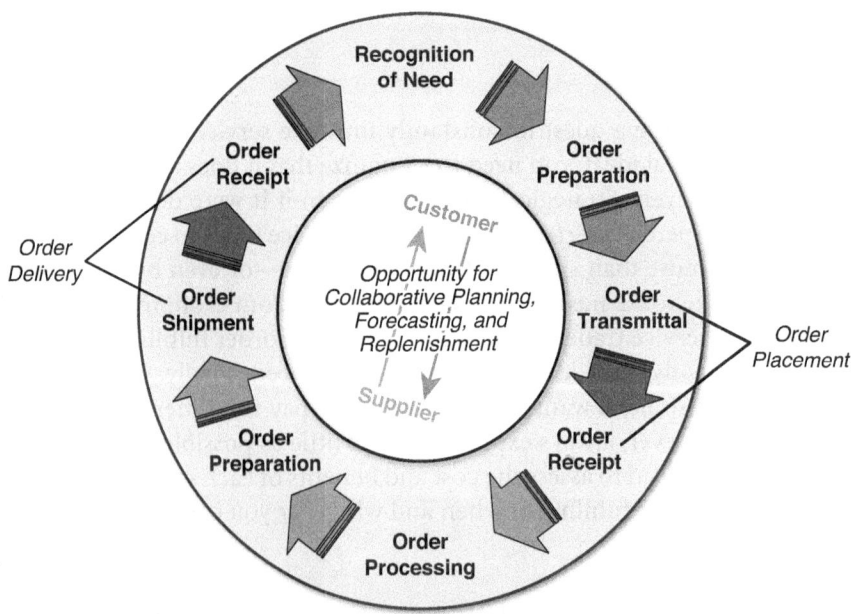

Figure 2-2   The complete order cycle

Mapping the Order Delivery Cycle: The SCOR Model

The following discussion details that portion of the order cycle that you as a supplier must design and manage to achieve outstanding delivery performance, which begins with order placement and concludes with order delivery. The Supply Chain Council has mapped this process as part of its Supply Chain Operations Reference (SCOR) model. Importantly, a process reference model does the following:

- Captures the "as-is" state and defines the desired future "to-be" state of a process
- Benchmarks operational details and performance of similar (i.e., reference) processes to establish best-in-class targets
- Describes the management practices, including performance measurement, personnel training, and technology implementation that lead to world-class performance[14]

The SCOR delivery process is widely considered as an exemplar for best-in-class operations. The SCOR model defines four distinct delivery processes: D1—Deliver Stocked Product, D2—Deliver Made-to-Order Product, D3—Deliver Engineered-to-Order Product, and D4—Deliver Retail Product. Figure 2-3 depicts the 15 activities in process D1—Deliver Stocked Product.

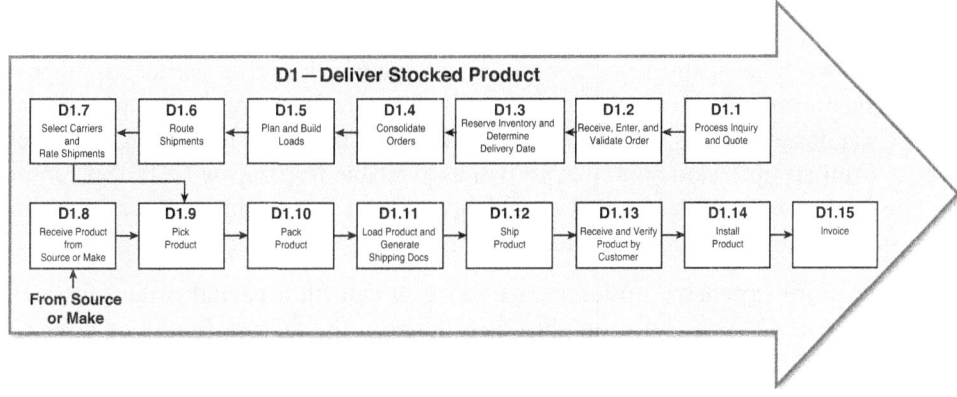

Figure 2-3   The SCOR delivery process[15]

D1.1—Process Inquiry and Quote

In the mind of the customer, the order cycle is initiated when you receive a request for information or other inquiry from a potential customer. The customer reaches out to you—and potentially other suppliers—for product information (e.g., specifications, pricing, and availability). Your job is to respond quickly with accurate, up-to-date information. A great product/service supported by rapid response helps you win orders.

D1.2—Receive, Enter, and Validate Order

If the customer decides to buy from you, she transmits an order, which you receive and process. In the past, this was done manually, often via fax or by phone as the customer talked to a customer service representative. Orders had to be manually entered into the supplier's system—a process that created many errors. The introduction of electronic data interchange enabled direct transfer of information, reducing costs, time, and errors. Today, many orders are placed directly via the Internet, enabling leading suppliers to provide customers with real-time order visibility.

D1.3—Reserve Inventory and Determine Delivery Date

As orders come in, customers either set the required delivery date or expect to receive from you a promised delivery schedule. If product is already inventoried in your local DC (i.e., available to deliver or ATD), you reserve it for the customer and set the delivery date based on your knowledge of picking, packing, and shipping times.

If the product is not available from the nearest shipping point, you can (1) ship from a more-distant DC or (2) wait for the product to be delivered from a supplier or from one of your own manufacturing operations and then ship the order. You need to compare tradeoffs between costs and lead times to determine which option is most appropriate for any given order. For some customers, you may sacrifice profitability to be able to promise a quicker delivery. Promising a delivery date based on when you will receive product from an upstream source is known as available to promise (ATP). It requires that you know—with high levels of confidence—when the product will arrive at your DC. Confidence in your supplier's information and delivery capabilities is critical.

Two other more-expensive options exist: (1) You can fill a partial order with existing inventory or (2) you can drop-ship directly from the factory (yours or your supplier's). Shipping partial orders increases handling costs for both buyer and seller and can diminish a customer's confidence in you. However, if the customer is about to stock out and thus lose customers or risk a production shutdown, a partial order may be the less-expensive option. When this occurs, you need to be able to promise delivery times for both the partial shipment and remaining items. Shipping directly from a supplier often increases coordination and shipping costs. Such shipments tend to be less-than-truckload or parcel and often involve longer distances.

D1.4—Consolidate Orders

After delivery dates have been set, you will want to check to see if you can leverage consolidation-driven cost savings. Most consolidation opportunities are found in the picking process (items to be picked for two separate orders are located close to each other in the warehouse) and in freight consolidation (two less-than-truckload orders are being shipped to similar locations). Efforts to consolidate introduce tradeoffs, which you must carefully evaluate. For instance, how much order cycle lead time are you willing to sacrifice to take advantage of consolidation efficiencies? A deep understanding of individual customer needs coupled with strong data analytics can help you make the right consolidation decisions and still promise competitive delivery times.

D1.5—Plan and Build Loads

Using the decision information from D1.3 and D1.4, you build a transportation plan, assigning individual orders to a specific carrier, route, and vehicle. A transportation plan answers three core questions: (1) What are we going to ship in each load? (2) Who is going to move the order? and (3) Which route will be used? The goal is to optimize transportation efficiencies and meet promised delivery times. Transportation planning is a very complex and time-consuming combinatorial process of matching orders to a variety of transport modes, carriers, vehicle capacities, and routes. Most companies use transportation management systems (TMS) to achieve more optimal loads and transportation plans.

D1.6—Route Shipments

As the talking points for D1.5 suggest, to develop an efficient transportation plan, you assign a "load" (usually a transportation vehicle) to a specific route for customer delivery. This is typically done by the TMS concurrently with D1.5.

D1.7—Select Carriers and Rate Shipments

Also as suggested by the description for D1.5, building an efficient transportation plan requires that you select a specific carrier to move and deliver an order. Your goal is to minimize costs while meeting customer service obligations. This requires that you match shipment size and delivery promise to carrier type and shipment cost. Small orders with next-day or two-day delivery requirements are assigned to premium airfreight carriers like UPS, FedEx, and DHL. Larger orders (single or consolidated) are assigned to less-than-truckload (LTL) or truckload service providers. You can minimize delivery costs by using TL carriers.

D1.8—Receive Product from Source or Make

Before an order can be filled, you must receive the product from an internal (factory) or external (supplier) source. As inbound shipments arrive at your DC, the order management system verifies their disposition. If the shipment contains items that are part of the ATP order, the product is immediately joined with the on-hand inventory to complete the order and prepare it for picking, packing, and shipping. If no outstanding order exists, the product is moved to storage. Informed and efficient receiving keeps costs down and allows you to make ATP delivery commitments.

D1.9—Pick Product

Based on delivery promises and the transportation plan, you develop a picking schedule so that the right product is picked from the DC in an accurate, efficient, and timely manner. Picking errors are costly. Most companies use radio frequency (RF) and bar codes or other technology (e.g., pick to light) systems to improve the picking process. Supported by appropriate training and measurement, these technology systems have greatly reduced picking errors.

D1.10—Pack Product

Once the order is picked, you are ready to pack it for shipping. The two primary considerations are (1) to protect the product to avoid damage in transit and (2) to prepare it for economical loading. Packing for placement/stackability can greatly improve cube and truckload efficiencies. In a value-added scenario, customers may expect you to prepare store-ready displays of specially sequenced racks (see earlier Toyota discussion).

D1.11—Load Product and Generate Shipping Documents

The final physical activity that takes place within your DC is to load the vehicle. For truckload shipments, sequencing is not nearly as important as maximizing overall vehicle capacity utilization (weighing or cubing out). However, for multistop milk runs or LTL shipments, sequencing is critical. Product should be loaded in reverse order of the delivery schedule. The first items loaded in the nose or front of the trailer (or container) will be delivered last. The last order loaded will be the first delivered.

Before the vehicle leaves your facility, you will finalize document generation. For domestic shipments, the essential documents include the commercial invoice and packing slip, bills of lading, and manifest. For international shipments, you may also need customs clearance documents, certificate of origin, and dangerous goods declarations. With complete and accurate documentation, you can legally turn the order over to the carrier.

D1.12—Ship Product

Shipping involves the physical movement of the order from your DC to the customer's designated ship-to point (e.g., customer DC, manufacturing plant, or retail store). For domestic shipments, actual movement often represents a small portion of the overall order cycle. For nonairfreight international shipments, transportation is usually multimodal, involves multiple handoffs, and accounts for a majority of the order cycle.

D1.13—Receive and Verify Product by Customer

Once the shipment is delivered, the customer's receiving team will inspect the order—checking for quality and quantity—and reconcile documents. Traditionally, four documents were matched: purchase order (PO), supplier invoice, receiving report, and inspection report. Many companies have streamlined receiving and rely almost exclusively on the PO and your invoice. If a problem with either the order or the documentation is found, the customer will withhold payment until you correct the deficiency. The customer may also require that you provide a corrective action report documenting what you have done to make sure the same problem does not reoccur.

D1.14—Install Product

Some products—typically operating equipment and other heavy machinery—require installation before the order cycle is complete. Installation must be performed to agreed-upon specifications before the customer will clear your order and begin to process your invoice for payment.

D1.15—Invoice

Once the customer clears the order—including installation—she processes the invoice so that you can be paid. Many companies certify suppliers to streamline the receiving and invoicing process and shorten the order cycle. Ford pioneered this process in the 1990s, redesigning its accounts payable process to adopt "invoiceless processing."[16] When orders arrived from certified suppliers, a database was checked to make sure a corresponding order was expected. If record of the order was found, the order was accepted and accounts payable issued payment directly from the PO. If problems are discovered later (e.g., during product use), certification status is revoked until the supplier demonstrates renewed trustworthiness.

To summarize, let's review the basic anatomy of an order fulfillment system. Every order system consists of a set of identifiable value-added activities and three distinct flows: an information flow, a physical flow, and a financial flow.[17] Figure 2-4 breaks down the SCOR delivery process D1 into these flows, showing that the first seven activities compose the information flow, the next seven make up the physical flow, and the last activity conveys the financial flow. Evaluating the order fulfillment process from this perspective provides vital insight into the skills and infrastructure you need to build an outstanding order fulfillment capability. For example, we see the following:

- Information triggers fulfillment and enables efficiencies across the order process. To build a world-class fulfillment capability, you need appropriate IT systems (e.g., Transportation Management Systems (TMS) or Order Management Systems (OMS)).

- Physical activities need to be carried out in a meticulous and efficient manner. You need good people, accurate and aligned measures, and appropriate infrastructure to succeed.

- Although the financial flow appears perfunctory, you need to pay close attention to this last step in the order process. Timely cash flow—i.e., receipt of payment—is essential.

- Great order execution requires that decision makers in logistics, manufacturing, and supply all grasp and perform specific roles. Order fulfillment is a cross-functional capability.

- The customer, a supplier, and a logistics service provider perform critical roles. Expertise and investment in specialized equipment make all three indispensable members of the process team. Order fulfillment is an interorganizational capability.

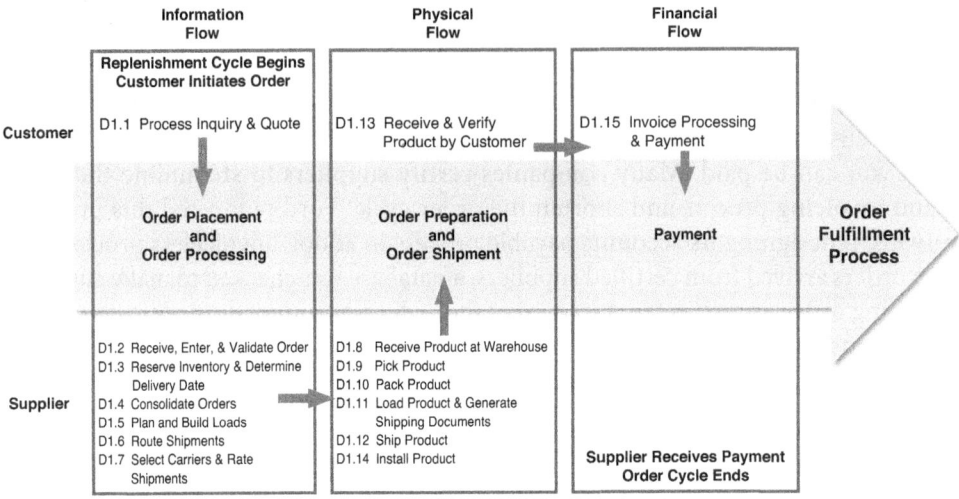

Figure 2-4 The anatomy of an order fulfillment system

Providing Postsales Customer Service

Although the order cycle is complete once you've delivered the order and received payment, postsales support is increasingly important in today's marketplace. Postsales service takes a variety of forms, including the following:

- **Technical information**—Many products sold today—from chemicals to computer systems—are extremely complex. Customers often need help figuring out how to use them correctly, especially when products fulfill multiple purposes. Packing a set of instructions with the product is insufficient to support customers. Web sites with answers to frequently asked questions or call centers staffed by experienced technical support representatives can help customers effectively use—and even maintain and repair—the products they buy.

- **Product support**—Over their usable life, many products need support beyond information. You may have experienced this with your laptop. As you spoke with the technical support person, you determined your laptop required a technician to fix it. Perhaps someone came to your house to repair your laptop. Alternatively, the laptop maker may have sent you a UPS packing box so you could ship the product in for repair. In many instances, your laptop was sent to a UPS logistics center where it was repaired and sent back to you. Getting product support right requires that you know the costs of alternative systems and compare them with the costs of poor service and customer dissatisfaction. Laptops and servers require very different support infrastructures.

- **Warranty**—Most high-value products come with a warranty, which is a promise that the supplier will support, repair, or replace its product for a specific period of time. Outstanding warranties can provide customers the comfort they need to try a product. For example, in 1998, Hyundai, a Korean carmaker with a reputation for iffy quality and reliability, decided to offer an industry-leading 10-year, 100,000-mile powertrain warranty. For several years, warranty costs were a loss leader, but they eased customer fears, increasing sales and growth. Over time, Hyundai improved its quality, reducing warranty costs and burnishing its reputation as a serious competitor in the auto industry.

- **Spare parts**—Some products, like heavy equipment and high-end computer servers, not only cost a lot to buy, but also incur very high costs when they break down and are out of service. A Caterpillar earthmover can cost thousands of dollars per hour when it is disabled; an IBM server can cost customers tens of thousands per minute. IBM promises four-hour response times.[18] Cat, which has equipment operating in some of the most distant and desolate places in the world, can ship customers the spare parts they need within 24 hours over 99 percent of the time.[19] These companies build extensive networks and robust processes to position spare parts and technical support so that they can get their equipment up and running as soon as possible. These networks are supported by parts management systems that provide excellent inventory visibility so that negative surprises are avoided.

- **Returns**—A few companies, especially bricks-and-mortar retailers, now use the returns process as a source of competitive advantage. A physical presence supported by no-hassle returns policies can make it easy for customers to return products. This convenience is something online retailers can seldom match. However, for returns to be a competitive weapon, a company must establish this as a part of its culture and policy. For instance, in recent years, customers have increased the frequency of returning products, raising company costs. To respond, many retailers tightened their returns policies. By contrast, Nordstrom and Costco have maintained their no-questions-asked, no-receipt-required returns policies. Customers often note these policies as a primary reason to shop at these retailers. Company lore at Nordstrom recounts a time when a clerk even gave a customer a refund for snow tires—even though Nordstrom never sold tires.[20] The fact that no one knows for sure whether the story is true or not attests to the power of Nordstrom's liberal returns policy.

In each of the preceding scenarios, operational excellence and customer satisfaction result as your order fulfillment design smartly integrates customer-oriented policy, information technology, empowered employees, and infrastructure into an effective postsales service capability.

The Cost of Order Fulfillment Failures

Think back to the thought exercise at the beginning of this chapter. You had just rushed to the store to buy an advertised item only to find the shelf empty. Aside from taking out your frustration on the unlucky customer service representative, what are your options? What are the costs of each option? Who incurs these costs? Table 2-1 lays out the basic answers to these questions.

Table 2-1   Comparing the Cost of Out of Stocks

Options	Costs
Postpone your purchase.	You are inconvenienced and must return at another time. Otherwise, the stockout does not drive any tangible costs.
Ask for a rain check and buy the item later at the sale price (in the B2B world, this is like a back order).	You are inconvenienced and must return at another time. The retailer's costs go up as it manages the rain-check process.
Buy a substitute product at the same retailer.	You may pay more for the substitute product. You end up settling for your second choice. The retailer's margins may be affected, depending on the comparative margins of the two products. The company that makes the out-of-stock product loses a sale—and the opportunity to create a positive second moment of truth.
Decide you really want the product and go to a different store to buy it.	Your search and acquisition costs go up. The retailer loses a sale (and maybe the value of other products you might have purchased if you had stayed in the store to shop). The company that makes the out-of-stock product retains the sale—but from another partner. This may affect margins.

Options	Costs
Be so ticked off that you swear you will never shop at that store again.	You may lose out on future deals available only at that retailer.
	The retailer loses the current sale as well as all of the purchases you would have made in future years.
	The company that makes the out-of-stock product may lose future sales if other retailers carry a smaller portfolio of its product line.
Be so ticked off that you tell everyone you know about your "horror" story.	The retailer loses your current sale and your future sales. Damage to the brand among your peer group (which may be very large in today's social networked world) may reduce sales.
	The company that makes the out-of-stock product may lose future sales if other retailers carry a smaller portfolio of its product line.

The Cost of Stockouts

Now, you may be wondering why you need to extend the original thought exercise and ideate potential customer options, calculating the costs associated with each one. The answer is simple: If you don't quantify the costs associated with stockouts, you cannot know how much money to invest in your order fulfillment capabilities. A quick perusal of Table 2-1 suggests that costs can (1) vary dramatically and (2) be very substantial. A closer look at Table 2-1 reveals that (1) calculating the real costs of stockouts is a messy process and (2) costs are not shared equally across members of the supply chain. These facts complicate the design of outstanding order fulfillment systems. Let's take a closer look at the three primary costs you are likely to encounter when a stockout occurs and then work through a sample scenario:

- **Back orders**—A back order occurs when (1) you do not have enough inventory available to fill a customer order and (2) the customer decides to keep her business with you. Typically, you ship what you have available and fill the rest of the order at a later time—once your inventory has been replenished. Although you may be pleased that the customer did not cancel the order and take her business to a competitor, back orders are inefficient. They increase your costs and reduce your profits. Specifically, each step in the delivery process (see Figure 2-3) must be repeated. Further, your ability to take advantage of scale economies in the delivery process is diminished as you pick, pack, and ship the smaller quantities created by splitting the original order. If the back order has to be expedited, per-unit costs can go up dramatically. Unfortunately, many companies have never quantified the true cost of a back order.

- **Lost sales**—Lost sales occur when the customer decides to take her business to a competitor. Unless you sell something for which there is no substitute, lost sales must be a concern—not just for the lost cash flow and profit, but also because you are giving your customer a reason to experience the competition's capabilities. The immediate cost of a lost sale is the profit associated with the sale. If the order was for 500 parts with a sales price of $50 each and a profit margin of 15 percent, the customer decision to buy from the competition will cost you $3,750 (500 units x $50/unit x 15% profit margin).

- **Lost customers**—If a customer decides the hassle of doing business with you is too great or if she decides she likes the product/service experience of buying from your rival, she will take her future business elsewhere. What does this cost? Few companies know. They have never done the homework needed to calculate the value of a loyal customer. Despite its importance, the concept of *lifetime stream of revenues/profits* is seldom used to design order fulfillment systems.

What is the true stockout cost? To figure this out, follow this three-step process:

Step 1: Estimate customer response—Because customer needs and circumstances are different, individual customers will react differently to a stockout scenario. This reality makes calculating the cost of a stockout difficult. As discussed previously, some customers will back order the out-of-stock product, others will buy from someone else, and some will permanently sever the business relationship. If you have great historical customer data, you may be able to estimate with accuracy how customers have responded to past stockouts and use this data to forecast future behavior. For this example, let's say 50 percent back order, 30 percent make a one-time buy elsewhere, and 20 percent remove you from the approved supplier list.

Step 2: Estimate consequence costs—Using customer and operating data, you need to calculate the costs incurred for each customer choice. For instance, the additional processing, labor, packing, and transportation costs associated with back orders could easily average $100 per back order. Using the scenario costs listed previously, we can assign a cost of $3,750 to the average lost sale. Finally, let's assume that you can forecast how often and how much a customer will buy from you in the future. This is extremely difficult to "guess" as product lines, technologies, competitors, and customer needs change over time. Using this forecast, you can calculate the net present value of the lifetime stream of profits associated with the permanent loss of a customer (let's say $50,000).

Step 3: Calculate the expected stockout cost—Using the information from steps 1 and 2, you calculate the expected cost of a stockout as follows:

Customer Response	Probability (i.e., Frequency)	Cost per Incident	Expected Cost
Back Order	.50	$100	$50
Lost Sale	.30	$3,750	$1,125
Lost Customer	.20	$50,000	$10,000
Expected Cost per Stockout			$11,175

Based on this analysis, you could improve overall financial performance by investing up to $11,175 in inventory (or some other remedy like better information) to eliminate a stockout of this item. Further, if you experience 100 stockouts per year, you can justify a $1,117,500 investment to improve your order fulfillment process to eliminate stockouts. As your efforts improve perceptions of your service, you may benefit from higher levels of satisfaction, loyalty, and lifetime stream of profits.

The Cost of Supply Chain Glitches

Supply chain glitches, including order fulfillment failures, can incur more wide-ranging costs than those described previously. For instance, event studies have quantified operational and stock price effects. Firms that experience and announce glitches report on average 6.92 percent lower sales growth, 10.66 percent higher growth in cost, and 13.88 percent higher growth in inventories.[21] Perhaps more important, glitches influence stock price valuations. An event study that looked at 838 glitch announcements found a 10.28 percent decrease in shareholder value.[22] The negative valuation effect is consistent across source as well as type of problem (see Figure 2-5). Put simply, it doesn't matter where the problem occurs or what the cause is, supply chain failures that undermine reliable delivery damage reputation and undermine stock price. The researchers who performed these studies argue that managers need to quantify these costs to make the "economic case for the major organisational changes that are needed to improve the reliability and responsiveness of supply chains."[23] They conclude,

> The fact that disruptions caused by external sources (supplier and customers) experienced a higher penalty suggests that these problems can be more expensive and time consuming for the firm to fix. This may be due to the firm's limited power to change their external partners' operations to solve the problems. This further underscores the need to form close and collaborative relationships with the various links in the supply chain. A firm must make sure that its supply chain partners see the value of working together to improve the performance of the supply chain network.[24]

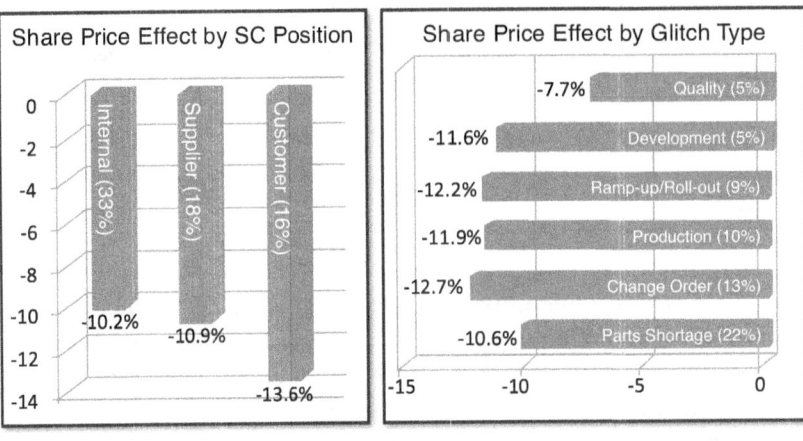

Figure 2-5 The stock-price-valuation effect of supply chain glitches

Conclusion

Establishing fast, flexible, robust, and efficient delivery processes to meet diverse customer needs is the key to order fulfillment system design. To do this, you need to define individual customers' needs and analyze numerous cost-service tradeoffs. Sometimes, however, delimiting all the variables to aptly define the "optimal" fulfillment system is a challenge. Consider Amazon.com's emergence as the world's largest online retailer with $61 billion in 2012 sales. Amazon began operations in 1995 as an online bookstore and went public on May 15, 1997. As a virtual retailer, Amazon was not constrained by physical footprint. It could offer customers far more titles than bricks-and-mortar rivals like Barnes & Noble. Almost immediately, Amazon became the face of virtual retailing, helping to popularize online shopping.

Analysts were thus dismayed when Amazon started to build large fulfillment centers. In 1997, Amazon built its second fulfillment center—a 202,000 square-foot facility in New Castle, Delaware. As a virtual retailer, Amazon wasn't supposed to need profit-diluting investments in bricks and mortar. Of note, such investments helped push Amazon's first profit out to the fourth quarter 2001—four years after its initial public offerings (IPO). Why did Amazon need to own and operate fulfillment centers? The answer lies in Jeff Bezos', Amazon's CEO, conviction that online shoppers desire rapid, reliable, low-cost delivery. In other words, customers want the back-office operations to be as smooth and hassle free as Amazon's state-of-the-art Web interface. As a pure virtual retailer, Amazon would always be a broker, offering breadth of product, but never able to deliver with the speed and dependability customers demanded.

From the beginning, Jeff Bezos wanted to change customers' buying habits—something Amazon could only do by building a world-class fulfillment capability. By 2013, Amazon operated 46 fulfillment centers in North America (37+ million square feet of warehouse space). Amazon had arguably become Walmart's most feared competitor.[25] Today's Amazon—armed with cutting-edge IT and an extensive distribution network—is close to achieving the elusive goal of same-day delivery.

What do you need to take away from this example? A myopic, short-term approach to tradeoff analysis and order fulfillment system design would have precluded Amazon's growth and long-term success. Order fulfillment design is part art and part science. You need to develop a deep understanding of customer needs and the costs of serving them. Chapter 3, "Developing a Winning Customer Fulfillment Strategy," addresses this challenge. You also need to establish a systematic approach to analyzing myriad design tradeoffs. Chapter 4, "Configuring the Network for Successful Fulfillment," explores this dilemma. You need to be able to discern when order fulfillment is an order qualifier and when it might be used to change the competitive rules. If you can use order fulfillment to change customer behavior and expectations, disadvantaging your competition, you may be able to justify investing in a world-class order fulfillment capability.

Endnotes

1. Federal Trade Commission. 1971. *Trade Regulation Rule Including a Statement of Its Basis and Purpose: Retail Food Store Advertising and Marketing Practices,* July 12.

2. Taylor, J., and Fawcett, S. 2001. "Retail on-Shelf Performance of Advertised Items: An Assessment of Supply Chain Effectiveness at the Point of Purchase." *Journal of Business Logistics* 22(1):73–89.

3. Corsten, D., and Gruen, T. 2005. *On-Shelf Availability: An Examination of the Extent, the Causes, and the Efforts to Address Retail Out-of-Stocks, in Consumer Driven Electronic Transformation.* Heidelberg: Springer, 131–149.

4. Taylor, J., Fawcett, S., and Jackson, G. 2004. "Catalogue Retailer In-Stock Performance: An Assessment of Customer Service Levels." *Journal of Business Logistics* 25(2):119–138.

5. Ibid 3; Holman, L., and Buzek, G. 2011. *Inventory Distortion—Retail's $800 Billion Problem.* Franklin, TN: IHL Group.

6. Harrington, L. (2007). "Change Drivers: Navigating the New Auto Supply Chain." *Inbound Logistics* (February). Retrieved June 28, 2012 from www.inboundlogistics.com/cms/article/change-drivers-navigating-the-new-auto-supply-chain/

7. Bowersox, D., Closs, D., and Cooper, M. 2012. *Supply Chain Logistics Management*. New York: McGraw-Hill/Irwin; Coyle, J. J., Langley, C. J., Gibson, B., and Novack, R. A. 2011. *Supply Chain Management: A Logistics Perspective*. Florence, KY: South-Western College Publications.

8. Stalk, G., Evans, P., and Schulman, L. E. 1992. "Competing on Capabilities: The New Rules of Corporate Strategy." *Harvard Business Review* 70(2):57–69; Stalk, G. J. 1990. *Competing on Time*. Cambridge, MA: Cambridge Press.

9. Ibid 7, Bowersox et al., 2012.

10. Ibid 7, Fawcett, S., Ellram, L., and Ogden, J. 2007. *Supply Chain Management: From Vision to Implementation*. Upper Saddle River, NJ: Prentice Hall.

11. Magnini, V. P., Ford, J. B., Markowski, E. D., and Honeycutt, E. D. J. 2007. "The Service Recovery Paradox: Justifiable Theory or Smoldering Myth?" *Journal of Services Marketing* 21(3):213–225; Michel, S., and Meuter, M. L. 2008. "The Service Recovery Paradox: True but Overrated?" *International Journal of Service Industry Management* 19(4):441–457.

12. Ibid

13. Juttner, U. 2005. "Supply Chain Risk Management." *International Journal of Logistics Management* 16(1):120–141; Smith, D. 2005. "Business (Not) as Usual: Crisis Management, Service Recovery and the Vulnerability of Organisations." *Journal of Services Marketing* 19(5):309–320.

14. Supply Chain Council, 2010. *Supply Chain Operations Reference (SCOR) Model, Overview—Version 10.0*. Cypress, TX: Supply Chain Council, Inc.

15. Ibid, 15

16. DeWitt, B., and Meyer, R. 2010. *Strategy Process, Content, Context an International Perspective*. 4th ed. Andover, Hampshire: South-Western Cengage Learning.

17. Ibid 10, Fawcett et al., 2007

18. Frei, F. X., Hajim, C., and Edmondson, A. 2002. *Dell Computers (a): Field Service for Corporate Clients* Boston, MA: Harvard Business School.

19. Glueck, J. J., Koudal, P., and Vaessen, W. 2006. "Putting a Premium on Service." *Supply Chain Management Review* 10(3):26–33.

20. Spector, R., and McCarthy, P. 1996. *The Nordstrom Way*. New York: John Wiley & Sons, Inc.

21. Hendricks, K., and Singhal, V. 2005. "Association between Supply Chain Glitches and Operating Performance." *Management Science* 51(5):695–711.

22. Hendricks, K., & Singhal, V. 2008. "The Effect of Supply Chain Disruptions on Shareholder Value." *Total Quality Management* 19(7/8):777–791.

23. Ibid, 23

24. Ibid, 23

25. MWPVL International Supply Chain, 2013. "Amazon.com Distribution Network." Retrieved on June 15, 2013, http://www.mwpvl.com/html/amazon_com.html

3

DEVELOPING A WINNING CUSTOMER FULFILLMENT STRATEGY

Opening Story: The Cost of Dropping a Baton

December 10

At 7:45, David walked into the conference room. He was 15 minutes early, but the entire team was already milling around the espresso machine. As David glanced around, he was startled to see Diane seated at the conference table. He hadn't expected she would join the team for the day's discussion, but his surprise passed quickly. Diane popped up at all sorts of meetings. David knew she wasn't there as a critic; rather, she was there to see how she could support the team's efforts. She had a knack for staying in touch with all of her team's key initiatives.

As team members settled into their seats, David greeted them with a hearty "Good Morning." Then he asked, "We are a couple of minutes early. While we wait for Doug to join us, do you mind if I share something I learned yesterday?" Seeing subtle nods of approval, David began: "Yesterday, I went to the state 5A track championships. My son and his 4X800 relay team earned the right to be called state champs. As I left, I thanked the coach and asked how the team was doing. His response surprised me. He said, 'Given the small team we brought, we're doing OK.' I asked, 'What do you mean small team?' He noted that several of the team's best performers were hurt. I asked about the girl's 4X100 relay team. He replied simply, 'They didn't qualify.' I was stunned. They dominated all season long. They were invincible. Seeing my dismay, the coach said, 'They were well out in front at conference, and then they dropped the baton. A season's worth of work evaporated in an instant.'"

"Two points quickly crystalized in my mind. First, timing IS everything! If they had dropped the baton at any other meet, they would probably be state champs today. Diane, do you remember what you said to me in the call that kicked off all of our efforts over the past couple of months?" Diane nodded affirmatively. David continued, "If my notes are correct, you said, 'If we are going to drop a ball, let's make sure it doesn't belong to Monster.' As a company, we are never going to be perfect, but we have to be perfect when it really counts."

"Second, competitive context matters! If my son had dropped the baton, he could have picked it up and run on. The 4X800 is forgiving. You can make up for a gaffe if your team is good enough. The 4X100 happens too fast. You just can't compensate for a glitch—much less a dropped baton. Now, 'Which race do you think we're running?'" Trina chimed in straightaway, "It's a 4X100!" David resumed, "That's the way I see it. In the old days, we could afford to deliver late from time to time—even to our biggest customers. Now, not so much! Rivals are too tough and customers like Monster are too demanding. If we drop the baton, we may not get a chance to pick it up. If we want to win tomorrow's order, we have to deliver on time today."

David's timing was perfect. Doug had walked in as David said, "...we have to deliver on time today." Naturally, Doug jibed, "I hope you've figured out how to do that!" "Good morning Doug," David responded with a smile. "Not yet, but we're making progress. We've mapped our as-is order fulfillment cycle and are meeting with customers to find out what they really expect from us. Trina and Lisé are going to share some thoughts about tailored logistics. We'd like to get your feedback on how well you think such a strategy might work. Lisé, you're on."

"Thanks, David, for setting the stage for us in such a fun way. If there is one thing we have learned in the past few weeks of digging into the profit and loss (P&L) impact of service failures, it is that it really does matter when we drop the baton. We have begun to document both top-line and bottom-line effects. Let me share a few high-level bullets and then pose a question to the team:

- First, we found evidence that it is less expensive to grow business with existing customers than it is to find and develop new customers. How big is the difference? We don't know—yet. But, we do know that we don't want to alienate customers once we have a foot in the door.
- Second, we discovered that no one really knows how much a customer is worth. How much money do we really make each time we sell to a customer? A lot of factors influence customer profitability, including what they order, when they order, and how they order. Let's face it, some customers are more costly—and more painful—to work for. Moreover, we've never calculated the lifetime stream

of profits for any of our customers. Honestly, it's tough to find a template for running the numbers. So, how much should we be willing to spend to meet a customer's needs? We don't know, but we need to find out.

- Third—and this is a critical point—as we have talked to customers, it is clear that they really do have different expectations. Most C-level customers know that they can't demand the same service as Monster. They've even pointed out that they don't need or want the same service. This really opens the door for tailored logistics. We don't have to be perfect with every customer, but we do need to deliver the right level of service to each customer.

- Finally, we've looked into the pros and cons of customer relationship management (CRM) software. Paul helped us with this. A CRM system could help us profile customer behavior and spend patterns so that we know which customers drive the most revenue. This would really help us answer the questions raised by the first three points. However, one of the downsides is that CRM systems are backward looking. They can't predict major shifts or inflection points in consumer behavior. They also can't tell us why customers do what they do.

With these points in mind, our question for the team is, 'How well prepared are we to segment our customers and develop the appropriate service menu for each customer?' Before you offer answers, let me break this down into three subquestions:

1. How well do we really know what our customers need and expect—on an individual customer-by-customer basis?
2. Do we grasp what their growth potential is—now and in the future?
3. Do we have reliable estimates on how much it costs to meet needs on an order-by-order basis?

We'd love to hear what you think."

Consider as you read:

1. How do the answers to Lisé's and Trina's four questions influence the value propositions and service experiences DWC should offer?
2. As an analyst, what information would you need to answer the questions posed by Lisé?
3. What approaches/tools can be used to segment customers for a tailored logistics strategy?

Developing a Winning Customer Fulfillment Strategy

> "When you can measure what you are speaking about, and express it in numbers, you know something about it; but when you cannot measure it...your knowledge is of a meager and unsatisfactory kind."
> —Lord Kelvin

In an ideal world, every company, including yours, would provide outstanding order fulfillment to every customer—all of the time. Unfortunately, you don't live and make decisions in a perfect world. Two factors make day-to-day order fulfillment messy: (1) uncertainty and (2) constrained resources. If you had perfect knowledge about what, how much, and when customers would order, you could produce exactly what they want and position it just in time where they will want it. Similarly, if you had unlimited resources (or if resources were free), you could almost always deliver to customer requests. However, you likely feel that you never have enough advance notice or sufficient resources to meet every customer request. You have to make your best guess about who will want what and then manage a variety of tradeoffs about how you will use resources to meet customer needs.[1] Fortunately, you are not alone: No company can provide the highest levels of customer service and order fulfillment to every customer all of the time.

Now, a little good news: Different customers have different needs and expectations. Intuitively, you know this. Some of your customers are extremely demanding. They expect perfection in every instance. Information's ubiquity has changed the power dynamic of buyer-supplier relations. Today's information-empowered customer has come to expect more than any previous generation of decision makers.[2] Customers now apply the Six Sigma goal of 3.4 defects per million to every aspect of operations, including logistics and order fulfillment.[3] However, even in this exacting environment, some customers simply do not need the exacting order fulfillment your most demanding customers require—especially if the higher service costs more. This reality sets up an important fulfillment system design principle and an associated corollary:

- **Design principle**—Not all customers are created equal.[4]
- **Corollary**—Not every order fulfillment experience is equally important.

What does this mean for you? Your challenge—and the key to long-term success—is to match your company's service offerings to each customer's real service requirements.

Managing Customer Relationships for Profitable Growth

Resource constraints mean you must figure out how closely to work with each of your company's hundreds, if not thousands, of customers. To develop a winning customer fulfillment strategy, you must be guided by the principle: Not all customer relationships are created equal—nor should they be.[5] Key accounts deserve intense managerial attention and resource dedication to achieve (1) high satisfaction levels and (2) profitable growth. Other customers—for example, one-time buys and purely transactional relationships—should be managed with an eye toward potential growth, but they may not value, need, or be worth unique order fulfillment efforts. Recognizing that different relationship types exist sets the stage to address two questions:

- What is the nature of the different relationship types shown in Figure 3-1's customer relationship continuum and how do these differences affect fulfillment strategy?
- Where should each customer be positioned on the customer relationship continuum?

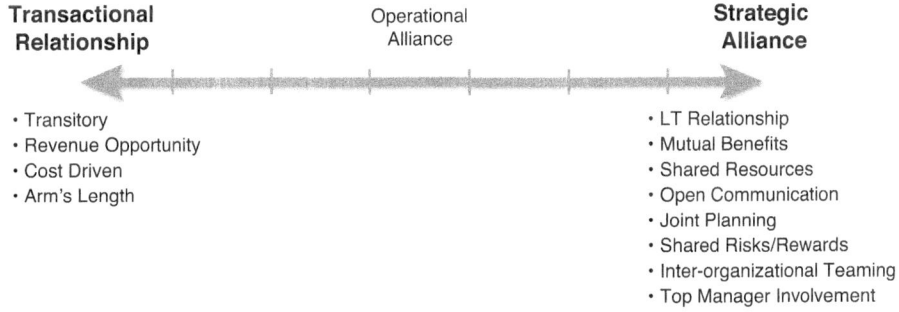

Figure 3-1   The customer relationship continuum

Managing Transactional Relationships

When you think about the nature of customer relationships, Pareto's law typically holds true: 80 percent of all customer relationships are transactional.[6] Transactional means transitory, cost driven, and arm's length. However, transactional relationships can represent vital revenue opportunities—now and in the future. Your goal for these nonstrategic relationships is twofold. First, strive for maximum efficiency. Customers value efficiency. Like you, customers are busy and do not have resources to waste with inefficient transactions, processes, or partners. Second, achieve good relations. Transactional customers almost always know they are not your most important account. Nevertheless, almost everyone wants to feel that you are treating her fairly.

Fair, efficient customer relations begin with a philosophy that says all customers should be treated in a way that supports a positive, productive arrangement. L. L. Bean's oft-copied customer service policy could be gainfully adopted and adapted by most companies (see Figure 3-2). Such a policy can help your firm become a "preferred" supplier to current—and future—customers. Managers should receive training in practices and behaviors that promote positive transactions and then be evaluated on their ability to cultivate proactive customer relations. Table 3-1 describes key practices that lead to good customer relations. You should apply each practice appropriately, depending on the potential strength, duration, and importance of a relationship.

L.L. Bean's Customer Service Policy

Customers are not dependent on us, we are dependent on them.

Customers are not an interruption of our work, but the purpose for it.

We are not doing our customers a favor by serving them, they are doing us a favor by giving us the opportunity to do so.

A customer is not someone to argue or match wits with. NOBODY ever won an argument with a customer.

Figure 3-2   L. L. Bean's customer service policy

Table 3-1   Practices That Can Turn Transactional Relationships into Growth Opportunities

Practice	Description
Confidentiality	You should maintain future technology road maps and other sensitive competitive information strictly confidential.
Equitable treatment	You should apply policies equally; that is, preferred status should be based on real tangible and transparent performance criteria (e.g., real profitability). Playing favorites creates mistrust and undermines relations.
Feedback	Periodic customer surveys or informal face-to-face dialogue can help improve the transaction process. You might ask, "What can we do to make your job easier?" and "How could we work together to reduce your costs?"
Integrity	You should always fulfill all contract obligations without hassle or argument. Model the integrity you expect to see from supply chain partners.
Mutual consideration	You should not unnecessarily burden customers. Quickly communicate when something goes wrong (e.g., a delivery delay) and openly discuss differences in opinion.

Practice	Description
Open communication	Good connectivity and face-to-face communication is critical to good relations. You need to learn to share all relevant decision-making information in a way that helps customers make good, timely decisions.
Personal contact	Personal relationships generate goodwill and reduce miscommunication.
Prompt response	You should handle customer suggestions as a valuable resource, evaluating them quickly and providing prompt feedback about how you plan to use them.
Training	Training is an investment in skills and relationships. You will find that it often makes sense to provide marketing and other support to customers.

What are the benefits of building fair and efficient relationships with transactional customers? For starters, fair and efficient relations help keep negotiations simple and facilitate problem resolution. Customers are also more forgiving of the occasional service failure if they view you as a preferred supplier. More important, good relations also open the door for future business. Never forget that it is difficult to forecast which relationships will be important down the road. By contrast, poor relations increase complaints—especially in a social media environment—damaging your company's industry reputation. Poor relations can also take a solvable dispute to litigation, costing your company time and money. To summarize, the benefits of fair and efficient relations are as follows:

- Reduced administrative costs
- Fewer complaints and a better industry reputation
- An opportunity to collect better competitive intelligence
- Higher profitability
- Future growth

Managing Strategic Alliances

True strategic alliances are rare and likely represent fewer than 5 percent to 10 percent of your company's customer relationships.[7] Customer alliances, however, often drive large revenue streams and represent key opportunities to develop new products and services and otherwise work with customers collaboratively to change the competitive landscape. Customer alliances are characterized by a long-term focus and intense interaction. Top management spends time and money to ensure these relationships are built on a solid foundation and for the long haul. For example, senior leadership commonly participates proactively in relationship development—and if needed, problem resolution. Cross-organizational teams are widely employed. A variety of resources are also shared, including engineering talent, training, financing, and operating capacity. Likewise, the risks and rewards of collaboration are shared. Trust promotes open communication and linked information systems enable strategic initiatives and day-to-day operations.

Two primary issues complicate alliance development: (1) determining which relationships merit this intensity of collaboration and (2) establishing the governance and operating routines to support everyday planning and execution. Research suggests that modeling alliance development as a three-phase process can help mitigate the downside of these challenges (see Figure 3-3).[8] Table 3-2 identifies and describes essential phase-appropriate practices that promote and sustain winning strategic alliances.

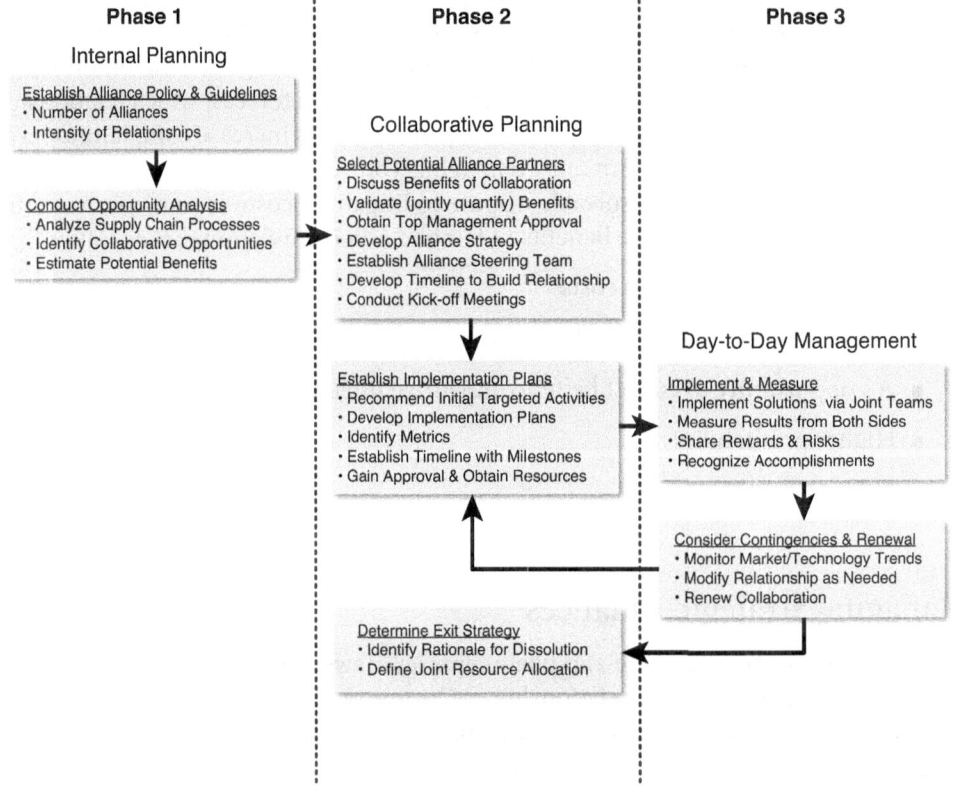

Figure 3-3   The alliance development process

Table 3-2 Alliance Management Practices

Practice	Description
Phase 1 Practices	
Alliance policy	A formal alliance policy provides guidelines to govern alliances. For example, policies denote who key contacts will be, how resources will be shared, and when investments will occur.
Screening template	A formal mechanism is used to identify, assess, and compare potential alliance partners. Should assess value co-creation potential and collaborative capabilities.
Phase 2 Practices	
Contract	Clear and concise long-term contracts govern successful alliances. Contracts run one to five years. Clearly communicated expectations, which are articulated in the contract, are often considered to be "the key" to alliance success.
Defined roles and responsibilities	Clear roles and responsibilities are defined and communicated. Explicitly stated roles and responsibilities reduce conflict and assure that key issues do not "fall between the cracks."
Confidentiality agreement	Confidentiality agreements protect competitive intelligence as well as proprietary technologies and processes. Such agreements should specify how any jointly developed technology is to be used in the future.
Continuous improvement clause	Continuous improvement clauses are standard in most alliances. Well-designed improvement clauses target cost, quality, delivery, and innovation performance and specify both rewards and penalties.
Exit criteria defined	Exit criteria are spelled out at the very beginning of the relationship. Even productive relationships can become one-sided or cease to be mutually beneficial. The long term almost never means forever.
Phase 3 Practices	
Dedicated teams	Dedicated teams are used to foster "personal" relationships and establish continuity between alliance partners, facilitating brainstorming and problem-solving initiatives.
Technology linkages	Technology linkages enable routine information exchange. A policy and culture of frequent, honest, and open information sharing support technologies that connect partners.
Problem resolution	A problem-resolution methodology is in place and clearly documented. It is used to resolve the misunderstandings that occur in even the best relationships.
Risk and reward sharing	Risks and rewards are shared on a mutually acceptable basis. Synergy requires that both sides of an alliance benefit from the relationship.
Aligned measures	Aligned and consistent measures help alliance partners evaluate their own and each other's performance. They also help identify problems before they become crises.

Phase 1: Internal Planning

The alliance-building process begins as you ask, "What should our alliance strategy look like?" To answer this question, you must assess whether or not closer customer relationships can really improve your competitiveness, market share, and/or financial performance. As hinted previously, great alliances don't just happen. Winning alliances are resource intensive and tough to manage. You have to carefully ascertain if (1) real value co-creation potential exists and (2) your company possesses the mind-set, skills, and patience to build successful alliances. Your alliance policy and guidelines should provide a framework to decide where customers belong on the relationship continuum as well as how many resources are available to dedicate to build stronger customer relationships.

With an alliance strategy in place, you are ready to conduct an opportunity analysis, which identifies (i.e., screens and evaluates) the right customers with whom to form an alliance. Be careful not to fall into the trap of trying to build too many alliances—especially with customers who offer no opportunity to change competitive dynamics via unique value co-creation. Rather, target specific opportunities with key accounts and then establish well-defined outcomes. Being specific will promote open communication, set expectations, and define roles and responsibilities. This is a critical part of quantifying the attainable benefits to be gained from the closer working relationship. Opportunity analysis is generally tedious and time consuming. It requires careful and detailed analysis of revenues and costs associated with the business relationship. However, if you have built fair and efficient relationships, customers will be willing to participate in the opportunity analysis.

Phase 2: Collaborative Planning

Once you have identified and prioritized customer opportunities, you need to take a joint deep dive into the analysis to document the expected value of the relationship—most likely through a joint-planning team. Your goal is to quantify the net gains/costs of the alliance to both companies. You also need to define how collaborative value co-creation (including higher levels of order fulfillment) will change the relationship as well as who will do what to ensure the highest level of strategic and operating efficiency. Specifically, what investments are needed, how much time and how many resources will each partner provide, and what benefits will emerge and how will you share them. As you validate the alliance as viable and profitable for both partners, you will be able to obtain top-management commitment.

Once the decision to move forward is made, you will put in place an alliance steering team and establish an execution timeline. You may want to use a kick-off meeting/celebration to signal that the deeper relationship is being developed. The alliance steering team then goes to work to nail down the nitty-gritty details that change the way two companies work together. Such details are tied to specific operational improvement activities and focus on goal identification, role specification, and the adoption of performance metrics and milestones. Minimum performance expectations are established and clearly articulated.

The final step in the initial planning process is to evaluate *"what-if"* scenarios. This analysis looks forward to consider (1) how changes in the marketplace might affect the relationship, (2) how shared resources/jointly developed capabilities might be managed, and (3) under what circumstances the relationship might be dissolved. Many managers mistakenly fail to consider the exit strategy. However, you need to realize that over time, leadership priorities will evolve and the competitive environment will change. The alliance may also fail to deliver on its initial promise. In each of these scenarios, the alliance may become obsolete or even dysfunctional. Planning the exit strategy up front—i.e., as part of the alliance definition process—will help you maintain a productive relationship or end the relationship amicably. Both outcomes will protect your reputation as a good partner to work with.

Phase 3: Day-to-Day Management

Once the relationship is set up, the steering committee transitions to an account management team with responsibility for execution, measurement, and renewal. What skills do you need? At the top of the list are good communication, team building, and process/project management skills. Because innovative alliances always run into stumbling blocks, patience is also needed. Table 3-3 lists additional intangibles that facilitate alliance success. Further, and often overlooked, is the ability to recognize and celebrate outstanding results. Great companies like Honda and Harley-Davidson make celebration part of their relationship-development programs.[9] Celebrating not only affirms people's hard work, but it also helps disseminate success stories, which build momentum for relationship transformation and future collaboration.[10]

Table 3-3 Intangible Attributes That Facilitate Alliance Success

Collaborative/joint efforts	Patience and perseverance
Collaborative continuous improvement	Personal relationships
Collaborative creativity and idea generation	Shared vision and objectives
Cultural fit	Trust
Mutual commitment to the relationship	Understanding of each other's businesses
Mutual dependence	Willingness to be flexible and tailor services

To keep the alliance fresh and productive, you must continually evaluate the alliance's performance and actively look for renewal opportunities. Scanning is vital to identify emerging customer needs, competitor strategies, market conditions, political shifts, and technology trends. If you can identify inflection points early, you can ensure that the alliance remains relevant, capturing as much value as possible. Complacency, by contrast, impedes long-term development as value-creation opportunities are missed. For example, despite years of working with Toyota on a variety of collaborative ventures, GM failed to join Toyota in investing in hybrid engines, essentially ceding the "green"-car market to Toyota's Prius. This failure by GM to actively pursue an emerging opportunity had negative long-term market share, revenue, and reputation costs.[11]

What are the benefits of building strategic customer alliances? Customers benefit from lower costs, decreased order fulfillment times, enhanced responsiveness, better product quality, and tailored innovation. Suppliers benefit from relationship stability—a key driver of steady revenue and better planning information. Specifically, longer-term, larger-volume contracts help suppliers plan capacity utilization (e.g., production, warehouse, and rolling stock), which helps reduce costs and improve service performance. Suppliers can also more easily justify investments in new product and process technologies. Further, both partners may gain access to each other's expertise and resources. For instance, Procter & Gamble (P&G) has provided smaller distributers capital to upgrade the plant and equipment to improve operating efficiency.[12] General Electric provides key customers Six Sigma training via a program called "At the Customer, For the Customer."[13] Honda provides similar expertise upstream to suppliers.[14] Such resource sharing is designed to improve both operating performance and relationship quality. The ultimate goal is to drive learning, create unique capabilities, and become partners in profit.

Relationship Takeaways for Fulfillment Strategy Design

Great customer relationships—both transaction and alliance—are hard to cultivate. The challenge begins with defining relationship intensity and continues through establishing the policies and systems needed to fulfill immediate customer needs while working to maximize long-term value co-creation. You should take the following five learning points away from the preceding discussion:

- **Stress appropriateness**—The discussion regarding relationship intensity should focus on the word appropriate. Many very good companies such as P&G have been caught in the mental-model trap that supply chain management (SCM) means close working relationships. As a result, they have invested scarce resources to build tightly coupled relationships with other members of the supply chain where no real advantage could be gained. That is, no real opportunity existed to co-create distinctive value. Establishing the wrong relationships with the right customers and suppliers almost always leads to disappointing return on investment (i.e., wasted investment or lost opportunity).

You need to think in terms of relationship appropriateness, which then defines relationship intensity. Unfortunately, very few companies specifically assess value co-creation potential and collaborative capability in the supplier selection/customer development process. Generally accepted metrics and methods to assess these performance dimensions do not exist. Thus, any manager who includes such criteria in the selection decision takes a serious risk. For example, at Bob Evans, a restaurant chain, a supplier selection decision was being made. The team purposely included a measure of collaborative capability in the process. Doing so changed the outcome, leading the team to select a much more expensive supplier than otherwise would have been chosen. Afterward, the managers shared that their biggest fear was that the desired collaboration might not emerge or they might not be able to document the value of the competitive benefits. If that happened, how could they possibly justify leaving so much money on the table? The good news: The collaboration worked and the managers were able to show the skeptics the money. But, in today's decision-making environment, where short-term results are demanded, how many managers are willing to take such a risk?

- **Look to the future**—As you assess and define relationship intensity, you need to carefully consider that today's decisions influence tomorrow's relationships. For example, some years ago, the chef of a small restaurant placed an urgent order for two boxes of peas with an Italian frozen foods distributor. The distributor prided itself on its highly responsive delivery capability. Indeed, it had built and promoted its ability to accommodate tough requests on short notice. However, on this occasion, the distributor refused the order. The chef later became the food and beverage manager of Italy's leading hotel chain. In his new position—one from which he could have ordered thousands of boxes of peas (among other things)—he refused to buy anything from the distributor.[15] The reality is that companies like Dell and Microsoft as well as Hewlett-Packard and Walmart started out in a garage or as a five-and-dime operation. You never know which of today's "C" customers will develop tomorrow's hit product or breakthrough technology. Thus, you really do need to establish the policies and build the processes that enable you to provide even C customers fair and efficient service. Enough of your competitors will fail to do so that when your customers grow, you will be a supplier of choice and grow with them.

- **Emphasize profitable growth**—One of your most important goals is to grow the top line—profitably. This can be done in two ways: Acquire new customers or retain and increase sales to existing customers. Research has shown that retaining customers is often the more profitable approach. It is often said that it costs five times more to make sales to a new customer than to an existing customer.[16] Others argue the cost ratio is higher—eight to ten times costlier to acquire new customers.[17] Some argue that some of the proclaimed benefits are illusory.[18] The bottom line: It's really hard to know exactly what the ratio looks like and it probably varies by industry and by company, but it is safe to say that it is indeed more profitable to grow sales with *satisfied* and *loyal* existing customers than to go out and convince new companies to become customers.

Truly satisfied customers emerge when you truly meet their needs. Thus, you have already sold them on your value proposition and your ability to deliver to promise. They pay less attention to your rivals, are less price sensitive, and are likely to buy more from you as long as you continue to perform. As a rule, it also costs less to serve them—relationships are already established and systems linked. Finally, satisfied customers generate positive word of mouth, which can bring in new customers. Consider this phenomenon carefully. Customers often talk to each other. When customers grumble about poor-performing suppliers, your satisfied customer is likely to drop your company's name as an alternative. By contrast, if your service is not good enough to retain customers, it probably is not good enough to attract new customers in a cost-effective manner. Perry Marshall summarized this reality, saying that companies that offer inferior customer service "only know how to replace angry customers with ignorant ones."[19]

- **Measure lifetime customer value**—How much is a satisfied and loyal customer worth? For most companies, this is an unanswered question. Consider the following anecdote. A supply chain analyst visited Japan to learn more about the nature of the Japanese kieretsu (buyer/supplier network). As he interviewed the owner of a manufacturer, he became intrigued by the intimate nature of the relationship. He asked, "How long have you been doing business with this customer?" The business owner responded, "My ancestors first started selling to this customer back in 1062." Now, despite the fact that not all customer relationships will, or should, endure 1,000 years, the vital question emerges, "What is a profitable customer worth over a lifetime?" You might ask this question a little differently, "What level of order fulfillment would you deliver to keep a profitable customer for 10 years? 100 years? 1,000 years?"[20]

A few companies have done the analysis to answer these questions with certainty. For example, Max & Erma's, an Ohio-based restaurant, became concerned that its employees viewed customers as a $20 meal or a $5 tip. In other words, service personnel might not understand how important customer service really was. To help senior decision makers determine how much to invest in improving customer service, the decision was made to crunch the numbers. How often did Max & Erma's best customers visit? How large was the average ticket? How many new, potential customers did these loyal customers introduce to the restaurant? What did the leadership team at Max & Erma's learn? Its best customers were worth $25,000 profit over a lifetime (that's a lot of burgers).[21] Knowing what a loyal customer is really worth can change the way you think about customers and what you are willing to do to create the loyalty needed to keep customers coming back.

- **Remember empathy**—Remember the old Swiss saying: "You cannot slice a piece of cheese so thinly that it only has one side." Every buyer/supplier interaction is seen from two viewpoints: yours and the customer's. You need to consider how your customers perceive each interaction as well as the overall relationship. You should likewise be aware that the following behaviors undermine the best-intentioned alliance strategies:

  - Excessive reliance on power to *motivate* cooperation: Partners view this as a coercive approach to extracting concessions and relationship value.
  - Failure to accurately account for and share risks/rewards: Identifying and quantifying risks and rewards are extremely difficult. You need to establish a transparent process and then follow through.
  - A short-term focus: This is manifest in a variety of behaviors, including (1) expropriation of the alliance benefits to maximize immediate profits and (2) treating partners as if they are only as good as their last performance.

A little empathy goes a long way toward building relationships that last—and that deliver both value and profits.

Segmentation Tools and Techniques

The essence of a tailored logistics strategy is to provide different customers the service they need and are willing to pay for. Executed well, tailored logistics can drive profitable growth.[22] Customizing service to meet each customer's specific needs, however, is far too complicated and expensive—and it really isn't needed or valued by the majority of your customers. Most companies, therefore, segment their customers and design service menus for each segment. Segmentation is the primary tool for defining the *appropriate* type of relationship to build with individual customers. We discuss two approaches to segmentation in the following pages: ABC classification and customer profitability analysis.

ABC Classification

ABC classification relies on the Pareto principle—also called the 80/20 rule—to provide basic guidance for customer segmentation. The Pareto principle observes that 80 percent of benefits are generated by your best 20 percent of all customers. Figure 3-4 depicts this relationship graphically. Because a very select group among this 20 percent—typically about 5 percent of your total customer portfolio—are responsible for a disproportionate share of your sales, profitability, and growth, they are classified as "A" customers. These top customers deserve the most outstanding, most customized levels of service your company can provide. The other high-value customers are labeled "B" customers and receive very high levels of service. The remaining 80 percent of your customers—typically transactional customers—are known as "C" customers and receive high levels of standardized service.

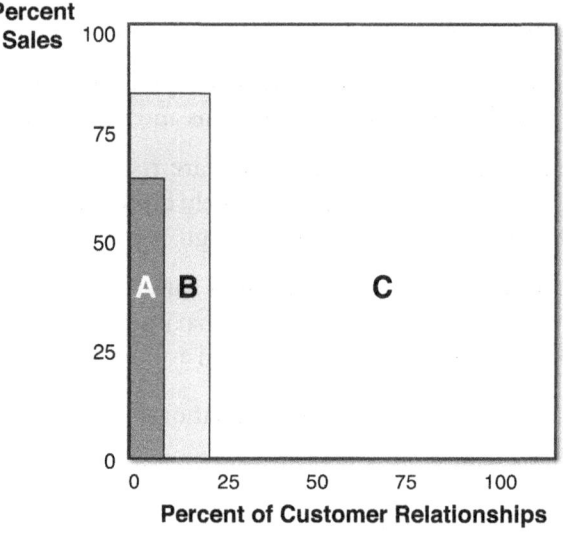

"A" customers are alliance candidates and receive the highest service levels.
"B" customers are important and should be managed with care, focusing on the future.
"C" customers should be managed efficiently with emphasis on fairness.

Figure 3-4   The Pareto principle applied to customer segmentation

Most companies perform "ABC" classification using sales numbers. Why? Because it is easy—that is, the data is readily available. The actual process typically takes place in two steps:

Step 1: Classify companies by sales—Table 3-4 shows how this is done using a sample data set for a pharmaceuticals supplier. Panel A lists the company's customers with sales figures in alphabetical order. Panel B shows the same data, but it is sorted on annual sales for the past year. The three largest customers account for 55 percent of the firm's sales. The next eight largest customers represent another 33 percent of total sales. The remaining 42 customers provide only 12 percent of total sales.

Table 3-4 Customer Segmentation Using ABC Classification

	Panel A: Alphabetical			Panel B: Sales			
	Customer	Sales		Customer	Sales	Class	% Sales
1	Abbott Laboratories	$5,111,000	1	Bristol-Myers Squibb	$26,100,000	A	55%
2	AbbVie	$265,400	2	Johnson & Johnson	$23,940,000	A	
3	Actavis Inc	$2,774,000	3	Merck & Co.	$15,130,000	A	
4	Aetna Inc	$5,753	4	Becton Dickinson	$9,867,000	B	33%
5	Agilent	$31,630	5	Amgen Inc	$7,553,000	B	
6	Alexion	$1,304,000	6	Abbott Laboratories	$5,111,000	B	
7	Allergan Inc	$1,754	7	CIGNA Corp.	$4,272,000	B	
8	AmerisourceBergen	$16,050	8	BIOGEN IDEC Inc.	$3,649,000	B	
9	Amgen Inc	$7,553,000	9	Pfizer Inc.	$3,258,000	B	
10	Bard (C.R.) Inc.	$199,700	10	Carefusion	$3,110,000	B	
11	Baxter International	$25,160	11	Actavis Inc	$2,774,000	B	
12	Becton Dickinson	$9,867,000	12	Cerner	$1,492,000	C	12%
13	BIOGEN IDEC Inc.	$3,649,000	13	Alexion	$1,304,000	C	
14	Boston Scientific	$983,600	14	WellPoint Inc.	$1,210,000	C	
15	Bristol-Myers Squibb	$26,100,000	15	Lilly (Eli) & Co.	$1,055,000	C	
16	Cardinal Health Inc.	$336,300	16	Boston Scientific	$983,600	C	
17	Carefusion	$3,110,000	17	Life Technologies	$909,300	C	
18	Celgene Corp.	$71,690	18	Waters Corporation	$796,900	C	
19	Cerner	$1,492,000	19	United Health Group	$597,900	C	
20	CIGNA Corp.	$4,272,000	20	Express Scripts	$561,400	C	

Panel A: Alphabetical			Panel B: Sales			
	Customer	Sales		Customer	Sales	Class	% Sales
21	Covidien plc	$30,480	21	Medtronic Inc.	$516,100	C	
22	DaVita Inc.	$158,400	22	Varian Medical	$410,100	C	
23	Dentsply International	$56,000	23	McKesson Corp.	$372,500	C	
24	Edwards Lifesciences	$50,360	24	Hospira Inc.	$342,900	C	
25	Express Scripts	$561,400	25	Cardinal Health Inc.	$336,300	C	
26	Forest Laboratories	$324,300	26	Forest Laboratories	$324,300	C	
27	Gilead Sciences	$320,600	27	Gilead Sciences	$320,600	C	
28	Hospira Inc.	$342,900	28	Stryker Corp.	$280,300	C	
29	Humana Inc.	$5,000	29	AbbVie	$265,400	C	
30	Intuitive Surgical Inc.	$220,400	30	Laboratory Corp.	$246,900	C	
31	Johnson & Johnson	$23,940,000	31	Intuitive Surgical	$220,400	C	
32	Laboratory Corp.	$246,900	32	Perrigo	$210,000	C	
33	Life Technologies	$909,300	33	Bard (C.R.) Inc.	$199,700	C	
34	Lilly (Eli) & Co.	$1,055,000	34	Tenet Healthcare	$194,100	C	
35	McKesson Corp.	$372,500	35	DaVita Inc.	$158,400	C	
36	Medtronic Inc.	$516,100	36	Regeneron	$112,400	C	
37	Merck & Co.	$15,130,000	37	Celgene Corp.	$71,690	C	
38	Mylan Inc.	$22,140	38	Dentsply International	$56,000	C	
39	Patterson Companies	$22,500	39	Quest Diagnostics	$55,780	C	
40	PerkinElmer	$39,550	40	Edwards Lifesciences	$50,360	C	
41	Perrigo	$210,000	41	PerkinElmer	$39,550	C	
42	Pfizer Inc.	$3,258,000	42	Agilent	$31,630	C	

Panel A: Alphabetical				Panel B: Sales			
	Customer	Sales			Customer	Sales	Class	% Sales
43	Quest Diagnostics	$55,780	43	Covidien plc	$30,480	C	
44	Regeneron	$112,400	44	Zimmer Holdings	$28,380	C	
45	St Jude Medical	$8,874	45	Baxter International	$25,160	C	
46	Stryker Corp.	$280,300	46	Patterson Companies	$22,500	C	
47	Tenet Healthcare	$194,100	47	Mylan Inc.	$22,140	C	
48	Thermo Fisher	$1,832	48	AmerisourceBergen	$16,050	C	
49	United Health Group	$597,900	49	St Jude Medical	$8,874	C	
50	Varian Medical	$410,100	50	Aetna Inc	$5,753	C	
51	Waters Corporation	$796,900	51	Humana Inc.	$5,000	C	
52	WellPoint Inc.	$1,210,000	52	Thermo Fisher	$1,832	C	
53	Zimmer Holdings	$28,380	53	Allergan Inc	$1,754	C	
	Total Sales	**$118,657,433**		**Total Sales**	**$118,657,433**		

Step 2: Modify classifications based on strategic issues—Because sales figures do not tell the entire story about how important customers are, you should evaluate the qualitative issues that might make a customer more or less important. You may want to consider the following:

- The customer represents a growing share of the company's sales.
- The customer possesses skills, technology, or another capability that will provide future market advantage.
- The customer controls scarce resources that will spur sales growth.
- The customer has unique access to downstream customers.
- Intensive collaboration may create market advantages: better quality, lower costs, shorter cycles, or unique service.

In the sample, these are the factors that would help define exactly where the lines are drawn between A and B customers as well as between B and C customers. Of course, if a customer like McKesson (currently a solid C customer) were to obtain a key product patent, it might be reclassified as a B—or maybe even an A customer.

You will find step 1 is pretty easy. Step 2, however, requires experience and judgment. You need to be an active scanner to identify and evaluate the relevant qualitative issues identified above. For example, it is difficult to assess whether closer collaboration might yield superior competitive advantage now and in the future. You need to perform an evaluation similar to the opportunity analysis desribed in the alliance development process (refer to Figure 3-3) to assess with confidence whether to reclassify a customer based on collaborative potential and capability. Finally, remember that relationships evolve, key technologies change, and essential capabilities emerge (sometimes in the most unlikely places). Therefore, you need to do the following:

1. Periodically reevaluate your classifications.
2. Manage all customer relationships with an eye to the future—even transaction-oriented, C-level relationships.

Your reputation as a proactive supplier can help grant access to the technologies and capabilities that will define tomorrow's market.

Customer Profitability Analysis

The goal of customer segmentation is to develop a profitable customer portfolio. Classifying customers based on sales is a good start, but you need to recognize that not all customers are equally profitable. Some customers buy a higher-margin portfolio of products and services. Others cost more to serve. A better—although more difficult and costly—way to segment customers is based on profitability. Classic customer profitability analysis often uses the following formula:

Gross Sales

− Returns

− Allowances

− COGS

Gross Margin

Unfortunately, because each customer interacts with your firm differently, this traditional cost-accounting approach—which allocates overhead to COGS using a standard basis—does not capture the true costs of serving individual customers. Activity-based costing (ABC) can help overcome this deficiency and provide better insight into actual customer profitability. The Chartered Institute of Management Accountants defines ABC as follows:

> An approach to the costing and monitoring of activities, which involves tracing resource consumption and costing final outputs. Resources are assigned to activities, and activities to cost objects based on consumption estimates.[23]

The key point is that activity-based costing ties actual costs to the customers who drive them.

How Does Activity-Based Costing Work?

Figure 3-5 depicts the logic of activity-based costing. Value is created and costs incurred as resources are used to produce products, deliver services, and meet customer needs.

To employ activity-based costing effectively, linking resource usage (i.e., costs) to customers, you need to be able to work backward from the end cost object, asking and answering four questions:

> **Question 1: What are our key cost objects?** Resources are purposely consumed—to create products or meet customer needs. In our current scenario, for which customers do you want to know the real cost of service? You probably want to begin your activity-based costing analysis with your most valued A customers.
>
> **Question 2: What process does the customer initiate?** Customers drive a variety of processes from new product development to production to reverse logistics. We are currently interested in the order fulfillment process.
>
> **Question 3: What activities make up the process?** Most value-added processes are complex. They are composed of a variety of activities that take place across organizational—and sometimes geographical—boundaries.

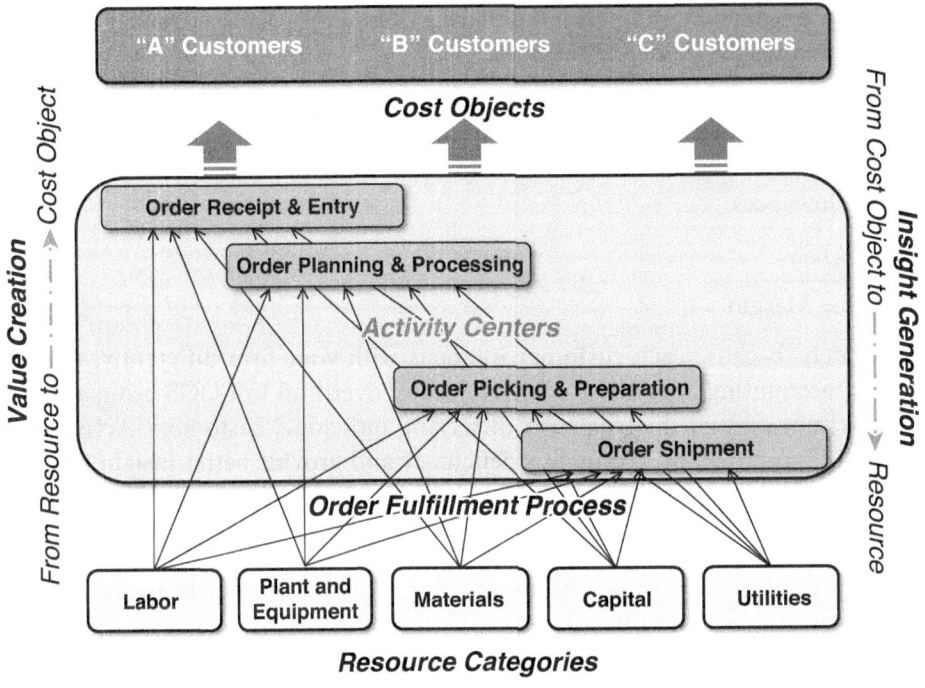

Figure 3-5 The logic of activity-based costing

Focusing on order fulfillment, think back to the SCOR delivery process (Figure 2-3 in Chapter 2, "Fulfilling Orders: The Nature of Modern Order Cycle Management"). It consisted of 15 distinct activities—a relatively simple and straightforward process. Even so, Figure 3-5 simplifies the process further by aggregating individual activities from the SCOR model into activity centers. Figure 3-6 disaggregates these activities and aligns them with the organizational departments that often have responsibility for their execution. Figure 3-6 highlights two important facts that complicate activity-based costing:

- **Level of aggregation**—Even the SCOR delivery process model aggregates activities in order to reduce complexity. For example, what are all of the specific steps involved in packing a product? Depending on product characteristics and customer needs, this could include palletizing, shrink-wrapping, labeling, staging, and loading. Each of the steps in the SCOR delivery process could be mapped and evaluated via time and motion studies. Because the accuracy is in the details, you need to define how much accuracy is really needed to design a winning fulfillment strategy and build strong customer relationships. More granular analysis drives up the costs of ABC.

- **Cross-functionality**—Because the activities that compose most value-added processes are housed across a variety of activity centers, departments, and functions, motivating cooperation to collect data can be difficult; that is, time consuming and costly.

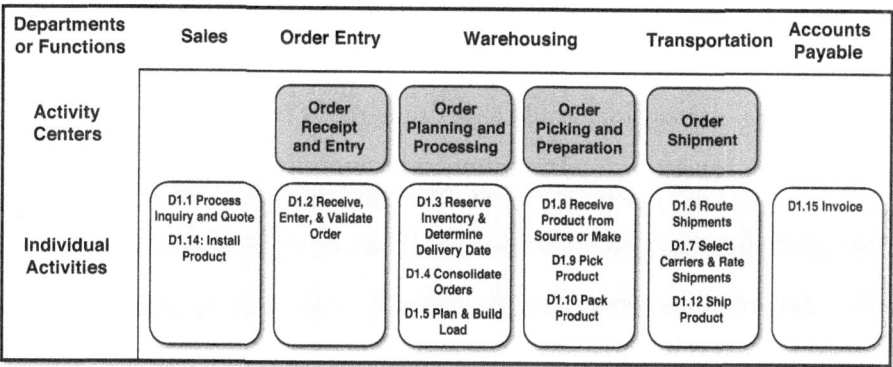

Figure 3-6   Tracing costs from cost objects to activities

Question 4: What resources are used by each activity? Performing the value-added activities shown in Figures 3-5 and 3-6 consumes resources, incurring costs to fulfill a customer order. A total factor productivity approach argues that you should identify and measure how much of each of five resources are required: labor, plant and equipment, materials, capital, and utilities. The real challenge is to tie each resource to a specific order. In the old days, workers needed to keep a log that identified when they started and ended work on a specific order. This manual approach made ABC tedious and expensive. Today, RFID and other sensor technology make tying individual resources to specific orders much more feasible.

The bottom line: Activity-based costing gives you an accurate and comprehensive view of the costs of service. In effect, ABC provides you individual customer P&L statements. But, tying costs to customers is more difficult and costly than traditional cost-allocation methods.

Why Activity-Based Costing?

Now that you know the basic mechanics of activity-based costing, let's discuss why you should invest in this more-refined approach to customer profitability analysis. In their quest to become customer-centric, some firms have adopted a "customer-delight-at-any-cost" strategy for their A customers. Only later did they find out they were losing money trying to keep their most demanding customers happy.[24] At one Fortune 500 company, for example, the adoption of an ABC methodology demonstrated that the firm's key accounts were unprofitable. However, this was discovered after the company had undertaken a seriously flawed strategy. To dedicate more resources to these supposedly critical customers, the firm had begun to stop selling to some of its low-volume C customers. The ABC methodology showed, however, that many of these C customers were the company's most profitable relationships. Simply stated, the C customers had been subsidizing the A customers. Without knowing it, the company had taken a purposeful, but wrong, strategy that would have dramatically reduced profitability.

How could a firm's biggest and "best" customers be unprofitable? To keep our discussion simple, let's return to the activities in the SCOR delivery process, as shown in Figure 3-5:

- Order receipt and entry
- Order planning and processing
- Order picking and preparation
- Order shipment

Some customers simply impose more costs for these different activities (see Table 3-5). Let's focus, for instance, on order picking and preparation. Customers can order product in a variety of pick quantities. A few of the options include the following:

- **Full pallet pick and ship**—Product is often received and stored in full-pallet quantities. On a cost/case basis, picking and shipping a full pallet of a single product is typically the most efficient and least costly.
- **Tier pick and pallet ship**—When a customer wants a smaller quantity, she may order a "tier," which is the number of cases that make up a single layer on a pallet. When tiers of different products are combined on a single pallet, a rainbow pallet results.
- **Case pick and pallet ship**—Individual cases of product are picked and stacked onto a pallet.
- **Case pick and case ship**—Individual cases are picked and either floor loaded or shipped via parcel delivery.
- **Each pick and each ship**—This option offers no labor or shipping economies and is typically the most expensive on a cost/case basis.[25]

Table 3-5 Customer-Imposed Costs in Order Fulfillment

Activity	Low-Cost Customer	High-Cost Customer
Order receipt and entry	Volume, blanket contract	Small orders placed one at a time
	Orders placed via Web portal	Orders placed by phone or fax
Order planning and processing	Steady ordering pattern	Sporadic ordering pattern
	Flexible delivery time windows	Tight time windows with late fees
Order picking and preparation	Full pallet picking	Case or each picking
	Pallet-stacked loads	Floor load
Order shipment	Full truckload orders	Less than truckload orders
	Infrequent expediting	Frequent expediting

Now, let's put this background to use. Using a traditional accounting process (total costs/total cases shipped), a Midwestern DC for a CPG company had determined that its cost per case was $.042. This number was used in customer pricing. The buyer for a mid-sized retailer—responding to a request from marketing—began requesting delivery of a combination of full pallets (40 percent) and tier and case (60 percent). After adopting an ABC-costing approach, the DC became aware that the costs per case for full pallets was only $0.015 compared with $.085 for tier and case shipments (a 567 percent difference). Fulfilling this customer's orders was 40 percent more costly than the DC managers had previously thought.

The following simplified exercise further exemplifies why you need to understand the costs of serving specific customers:

- OrderFill, a 3PL shipper, incurs $460,000 of costs each week. OrderFill's largest customer, H. S. Sponge, represents 25 percent of orders and sales. Closer analysis reveals that H. S. Sponge drives 15 percent of the order entry costs, 25 percent of the order planning and processing costs, 30 percent of the order picking costs, and 40 percent of the order shipping costs.

- What is your profit margin if H. S. Sponge has negotiated a weekly shipping price of $135,000?

Table 3-6 breaks down the costs and shows the margin impact of traditional allocation and ABC methods—which turn out to be very meaningful.

Table 3-6 Evaluating the Profitability of Customer Relationships[26]

Panel A: Traditional Overhead Allocation Based on Orders

Cost Categories	
Salaries	$240,000
Supplies	60,000
Depreciation	40,000
Overtime	30,000
Space	60,000
Other	<u>30,000</u>
Total	$460,000

As traditional costing allocates costs are based on the percent of orders H.S. Sponge represents, the costs of service are calculated as follows: .25 × $460,000 = $115,000

Profit margin: ($135,000 − $115,000) ÷ $135,000 = **14.81%**

Panel B: Activity-Based Costing

Cost Categories	
Order Receipt and Entry	$46,000
Order Planning and Processing	92,000
Order Picking and Preparation	138,000
Order Shipment	<u>184,000</u>
Total	$460,000

As activity-based costing is based on the costs of the activities required to meet a customer's needs, the costs of service are as follows:

$$.15 \times \$46,000$$
$$.25 \times \$92,000$$
$$.30 \times \$138,000$$
$$\underline{.40 \times \$184,000}$$
$$\$144,000$$

Profit margin: ($135,000 − $144,000) ÷ $135,000 = **−6.67%**

Adapted from Fawcett, Ellram, and Ogden, 2007

OrderFill goes from a healthy profit to a substantial loss. Because it doesn't make sense to lose money to keep customers happy, what can you do if you find your company in this situation? Three principal options exist:

Option 1: Collaborate to reduce the costs of service—One benefit of an ABC-costing approach is that it not only shows you what the real costs of service are, but it also reveals what activities/behaviors/decisions are driving costs. This insight enables you to enter into a fact-based dialogue about how you can work with customers to reduce the costs of service. You will find that customers often have no idea what some of their requests really cost. You may also find that your customers do not understand how some of their internal policies/processes affect costs of service. If they value you as a supplier, and want to avoid higher prices, they will be interested in working with you to bring costs down—once you show them the numbers.

Option 2: Raise prices—If you deliver best-in-class service, you will almost always be able to negotiate a price that yields a reasonable profit. The key is operational excellence. That is, your customer needs to know that other suppliers will either have higher costs or deliver inferior service. A second advantage of the fact-based dialogue enabled by ABC-costing is that you can show the customer that she is truly getting value for her money.

Option 3: Fire the customer—If a customer is not willing to collaborate to reduce the costs of service or to accept a reasonable price increase for valued services, then you may need to stop doing business with that customer. Accurate costing enables you to make this type of tough decision. Of course, before you fire a customer, you should make sure that you are not giving up other valuable benefits. For instance, some customers test out new suppliers with small quantity orders and potentially demanding service scenarios—the ideal recipe for an unprofitable relationship. However, if you pass the test, order quantities may increase even as the demanding test scenarios diminish.[27]

Tailored Logistics: The Right Service for Each Customer Segment

Once you understand both customer needs and potential as well as costs of service, you are ready to develop and articulate a tailored fulfillment strategy. Many companies classify their customers into three broad categories. Relationship strategies and associated service standards are defined for each customer category:

Category A: Customers of choice—Sometimes called key accounts, these are your best and most important A customers. They drive large revenue streams, represent good growth potential, are well served by and value your firm's product/service offerings, and are profitable. Investing resources to grow this business makes strategic sense. Consider the following relationship-investment possibilities:

- Communicate frequently and at multiple levels, including senior management.
- Establish cross-company teams to investigate growth opportunities, jointly reduce costs, and solve problems.

- Integrate information systems to enable efficient order fulfillment. Share real-time information on inventory levels, order status, and expected future demand as well as capacity constraints.
- Build processes to support exceptional service and accommodate special customer requests.
- Put in place policies to support extraordinary efforts to meet customers' unexpected needs.

Category B: High potential customers—These high-value customers (most likely all of your B and perhaps some of your A customers) merit close attention and great service, but not the same level of resource dedication. The truth is that "customer-of-choice" relationships are so resource intensive that a firm can usually only support a very small number of them. Consider the following policies and behaviors:

- Proactively seek customer input to meet emerging expectations. As the market changes, these customers may become customers of choice. To position your firm for growth, you want to be a supplier of choice now. This requires consistently high-service performance (see Table 3-7).
- Designate a point of contact (not a cross-company team) to achieve consistency and establish a personal touch.
- Link information systems to minimize transaction costs.
- Consider a long-term contract to clearly state expectations, roles and responsibilities, and performance metrics.

Category C: Transactional customers—These C-level customers receive very little personal touch; however, you cannot afford to take these customers for granted. As noted previously, these customers are often small but highly profitable—and they may grow into tomorrow's Apple or Walmart. High levels of reproducible service are the key (see Table 3-7).

Table 3-7   Segmented Service Offerings

Service Offering	Customers of Choice	High-Potential Customers	Transactional Customers
On-time delivery	98%	95%	90%
Lead time	Next day	48–72 hours	Within week
Time window compliance	15-minute	On day requested	On day promised
Fill rate	98%	95%	90%
Complete orders	95%	90%	85%
Payment terms	3/10 net 30	3/10 net 30	2/10 net 30
Customer support	Cross-company team	Designated POC	Web site/call center

Today, many leading-edge companies increase the level of service customization by offering a menu of ancillary service offerings. In effect, they offer a cost-of-service menu for services that some customers want to buy, but others do not. Pillsbury launched its tailored menu after implementing activity-based costing. Extra-cost options included modular promotional pallets, direct store delivery, cross-docking, and one-way pallets.[28] Of course, menu pricing could be used for any service that is emerging as important to your customer base, but has not yet become a standard expectation. Finally, the ultimate tailored logistics occurs in customer-of-choice relationships where one-of-a-kind service offerings are developed to support new value co-creation strategies. These service-offering incubators can act as the source of tomorrow's menu items.

Conclusion

To understand strategy is to understand why some companies win and others fail. Socrates once remarked to Nichomachides that the role of strategy is to plan the use of resources to achieve objectives—that is, to destroy the enemy.[29] In our context, your objective is to use your firm's constrained resources to fulfill customer needs. If you do this better than your competitors, you will grow your business with existing customers and steal your rival's customers (a surefire route to profitable growth—and victory).

From an order fulfillment perspective, knowing precisely what each customer wants and being able to deliver exactly to each customer's expectations is the key to success. It is order fulfillment's unattainable holy grail. The good news: "Big Data," print manufacturing, and other emerging technologies promise to someday move this ideal from the realm of science fiction to that of virtual reality. The bad news: For now, the closest you will get to perfect order fulfillment is a well-conceived and well-executed customer segmentation strategy. You won't be perfect all the time, but you won't drop the ball when it counts the most against you.

Ultimately, your ability to establish a winning order fulfillment strategy depends on how well you know your customers' needs, their potential for growth, and the costs of serving them. Recall Lord Kelvin's quote from the beginning of the chapter, "When you can measure what you are speaking about, and express it in numbers, you know something about it; but when you cannot measure it…your knowledge is of a meager and unsatisfactory kind." When you can express the value of each customer relationship in concrete and credible language, you will be able to define and build appropriate relationships and appropriate service offerings.

Developing a customer-segmentation fulfillment strategy is not easy—nor is it a one-time event. The world and customer expectations change too quickly. You will need to revisit your segmentation strategy and your menu offering on an iterative basis. A constantly and correctly evolving fulfillment capability can help you change customer perceptions and expectations—and thereby the competitive rules of engagement. You can use order fulfillment to disadvantage the competition. Think back to the last chapter's brief description of Caterpillar and IBM's spare parts and service fulfillment capabilities. Customers have come to rely on these suppliers of choice for unrivaled and dependable order fulfillment. For a rival to successfully enter the market, it must be able to either match Cat's and IBM's investment or find a way to make the spare parts/service game irrelevant (e.g., produce a product that never suffers a breakdown). So far, rivals have foundered in this effort. The bottom line: A winning order fulfillment strategy can get you into the game and it can keep rivals out.

Endnotes

1. Smith, L., Andraski, J., and Fawcett, S. 2011. "Integrated Business Planning: A Roadmap to Linking S&OP and CPFR." *Journal of Business Forecasting* 29(4):4–13.

2. Lucas, J. M. 2002. "The Essential Six Sigma." *Quality Progress.* (January):27–31.

3. Bowersox, D., Closs, D., and Cooper, M. 2012. *Supply Chain Logistics Management.* New York: McGraw-Hill/Irwin.

4. Fawcett, S., Ellram, L., and Ogden, J. 2007. *Supply Chain Management: From Vision to Implementation.* Upper Saddle River, NJ: Prentice Hall.

5. Ibid, 4

6. Duffy, D. 2005. "The Evolution of Customer Loyalty Strategy." *Journal of Consumer Marketing* 22(5):284–286.

7. Duffy, R., and Fearne, A. 2004. "The Impact of Supply Chain Partnerships on Supplier Performance." *International Journal of Logistics Management* 15(1):57–72.

8. Fawcett, S., Magnan, G., and McCarter, M. 2008. "Supply Chain Alliances and Social Dilemmas: Bridging the Barriers That Impede Collaboration." *International Journal of Procurement Management* 1(3):318–341.

9. Nelson, D., Mayo, R., and Moody, P. 1998. *Powered by Honda.* New York, NY: John Wiley and Sons, Inc.; Nelson, D., Moody, P., and Stegner, J. 2001. *The Purchasing Machine.* New York: The Free Press.

10. Fawcett, S. Andraski, J., Fawcett, A., and Magnan, G. 2009. "The Art of Supply Change Management." *Supply Chain Management Review* 13(8):18–25.

11. Muller, J. 2013. "How GM Lost—and Found—the Path to Innovation." *Forbes,* January 13. Retrieved September 18, 2013, from http://www.forbes.com/sites/joannmuller/2013/01/13/how-gm-lost-and-found-the-path-to-innovation/

12. Fawcett, S., and Magnan, G. 2002. "The Rhetoric and Reality of Supply Chain Integration." *International Journal of Physical Distribution and Logistics Management* 32(5):339–361.

13. Adrian, C. 2004. "A Loan and a Helping Hand." *CFO Magazine,* October 7. Retrieved September 17, 2013, from http://www.cfo.com/printable/article.cfm/3238461/c_3262722?f=options

14. Ibid, 9, Nelson et al.

15. Whybark, D. (1989). "Frisbee Frozen Foods." In *International Operations Management: A Selection of IMEDE Cases.* Ann Arbor, MI: BPI/Irwin.

16. Hart, C., Heskett, J., and Sasser, W. 1990. 'The Profitable Art of Service Recovery." *Harvard Business Review* 68(4):148–156; Peters, T. 1987. *Thriving on Chaos.* New York: Excel/A California Limited Partnership.

17. Unal, C. 2013. "Attracting Vs Retaining Customers—Get Your Priorities Straight!" *More Than Shipping,* Retrieved July 31, 2013, from http://morethanshipping.com/attracting-vs-retaining-customers-get-your-priorities-straight/

18. Keiningham, T., Vavra, T., Aksoy, L., and Wallard, H. 2005. *Loyalty Myths: Hyped Strategies That Will Put You Out of Business—And Proven Tactics That Really Work.* Hoboken, NJ: John Wiley & Sons, Inc.

19. Marshall, P. 2013. "Getting and Keeping Customers." *PerryMarshall.com,* Retrieved September 17, 2013, from http://www.perrymarshall.com/marketing/m17/

20. Ibid, 4

21. Blackwell, R. 1997. *From Mind to Market: Reinventing the Retail Supply Chain.* New York: Harper Business.

22. Fuller, J., O'Conor, J., and Rawlinson, R. 1993. "Tailored Logistics: The Next Advantage." *Harvard Business Review* 71:87–98.

23. Edwards, S., and Technical Information Service. 2008. *Activity Based Costing: Topic Gateway Series No. 1.* London: Chartered Institute of Management Accountants.

24. Bowersox, D., Calantone, R., Clinton, S., Closs, D., Cooper, M., Droge, C., Fawcett, S., Frankel, R., Frayer, D., Morash, E., Rinehart, L., and Schmitz, J. 1995. *World Class Logistics: The Challenge of Managing Continuous Change.* Oak Brook, IL: Council of Logistics Management; Fawcett, S., and Swenson, M. 1998. Customer Satisfaction from a Supply Chain Perspective: An Evolutionary Process in Enhancing Channel Relationships. *Journal of Consumer Satisfaction, Dissatisfaction and Complaining Behavior* 11:198–204.

25. Coyle, J., Langley, C., Gibson, B., and Novack, R. 2011. *Supply Chain Management: A Logistics Perspective.* Florence, KY: South-Western College Publications.

26. Ibid, 4

27. Kaplan, R., and Cooper, R. 1998. *Cost and Effect: Using Integrated Cost Systems to Drive Profitability and Performance.* Boston, MA: Harvard Business Review Press.

28. Ibid

29. Fielding, S. 1738. Xenophon's Memoirs of Socrates. Self-published. London.

4

CONFIGURING THE NETWORK FOR SUCCESSFUL FULFILLMENT

Opening Story: Moving Parts

February 14

Let's Join the Task Force Meeting in Progress:

David has been leading a discussion on the difficulties the team is encountering as it strives to reimagine DWC's order fulfillment infrastructure. David had asked the team to watch a 2-minute Honda television commercial called "The Cog." As David introduced the clip, he invited, "As you watch, consider how the development of this commercial is a metaphor for us and our efforts to create a world-class order fulfillment system. I look forward to your reaction."

As the ad concluded, Doug was quick to react, saying, "That was fascinating. I loved the special effects. But, I don't see how an effects-laden commercial of interacting auto parts applies to our order fulfillment problems." Trina swiftly responded, "Wait a minute Doug. They filmed the entire sequence without special effects. That's what makes it unique. And that's what makes it a model for us. *It is a true chain reaction!*" Lisé interrupted, saying, "Are you serious Trina? They did all of that without special effects? How'd they pull that off? If anything is off just a little bit, if any interaction fails to behave as designed, the whole thing breaks down." Anxious to push things along a little, Diane chimed in, "That's your point, David, isn't it? It's all about the interactions and how everything has to work together as a system. So, how did they do it?"

"OK, let me share the rest of the story. I think you will all see how this applies to us. In 1998, Honda's market share in Europe took a serious downturn. Honda's cars were viewed as staid—boring. In one survey, 25 percent of all potential customers said they, 'wouldn't dream of buying a Honda as their next car.'[1] Honda realized that it needed to reintroduce the brand. It needed to create a warm, consumer-friendly, and technologically savvy image.[2] The decision was made to show a Honda from the inside out, but in a unique, riveting way. The chain-reaction idea was born. Unfortunately, eye-catching and memorable interactions among moving auto parts create a less-predictable, hard-to-control chain reaction. Does that sound familiar?"

Paul took the bait, saying, "We certainly have discovered that our order fulfillment process acts as a chain reaction. With so many interdependencies, especially across departments, if a problem occurs anywhere, it can screw up performance for everyone." "Absolutely," David agreed. Lisé added, "The tradeoff dilemma also sounds familiar." Continuing, she noted, "Now that we are working on the 'to-be' design, we find that every improvement idea brings unexpected, sometimes surprising consequences for someone somewhere else in the order fulfillment process."

"That's a perfect segue to 'The Cog's' core message for us," David resumed. "To make the chain reaction work, the production team needed keen insight into the interactions among the moving parts. To develop this precise understanding, the director, Antoine Bardou-Jacquet, devoted two months to ideate and create hundreds of conceptual drawings of possible interactions.[3] The ideas that made it into the script were meticulously tested—one interaction at a time. Ideas that weren't viable were discarded. During this testing process, the production team learned that tiny deviations in things like settling dust, temperature, and humidity altered part movement enough to stop the chain reaction.[4] In the end, 606 takes were required to make 'The Cog.'[5] On the first day of filming, it took 90 minutes to get the first transmission bearing to roll into the second.[6] Nothing was merely intuitive. Nothing came easy."

"I get it, David! Your metaphor fits perfectly. Let me take a shot at making it a bit more concrete," volunteered Trina as she moved to the whiteboard. As she began to write, she said, "The 200 conceptual drawings are analogous to our rough-sketch 'to-be' map. They give a good idea of what could be possible, but they don't take reality fully into account. Only when the interactions were tested did the production crew learn how each part of the system really works. Just as Honda had to discard some ideas because they wouldn't work in real life, we have found unacceptable financial and performance tradeoffs associated with each of our ideas for redesigning our order fulfillment infrastructure. Because we don't know how all of the parts of our fulfillment system interact with each other and the rest of our value systems, we find real and substantial downsides with every decision we consider. If we don't develop the understanding that Honda gained through four months of testing, we will keep beating our heads against the law of unintended consequences. How'd I do, David?"

The COG	DWC's Order Fulfillment Redesign
200 conceptual drawings	Hypothetical "to-be" map
Ideas that would not work	Ideas with counterproductive tradeoffs
Four months testing = Intimate insight	??? = Lack of understanding

"Beautifully! The question is: What is our equivalent of four months of testing?"

Consider as you read:

1. Why are tradeoffs so prevalent and pervasive across supply chain systems, including order fulfillment?

2. As an analyst, what information would you need to make tradeoffs visible?

3. What approaches/tools can be used to mitigate the downside of unintended consequences in the order fulfillment design process?

Configuring the Network for Successful Fulfillment

"Education is the ability to perceive hidden connections between phenomena."
—Vaclav Havel[7]

Are you a fan of sports movies? If so, you may remember baseball manager Jimmy Dugan from the movie, *A League of Their Own*. At a particularly poignant moment in the movie, when star catcher Dottie (Gina Davis) is about to quit the team, saying, "It just got too hard," Dugan (Tom Hanks) responds, "It's supposed to be hard. If it wasn't hard, everyone would do it. *The hard* is what makes it great." Dugan could easily have been the keynote speaker at most firms' kickoff meeting for the reimagination of customer service and order fulfillment systems. You, just like everyone else, know that getting these systems right is vital to your company's long-term competitiveness and profitability. Even so, most companies really struggle to establish an order-winning delivery capability. You might ask, "Why?" The answer: Because it is really hard to design and develop the infrastructure to provide consistently great delivery and outstanding service. Along with deep knowledge of customer needs and great supply chain relationships, you need vision, sustained investment, and patience—uncommon traits in a world driven by short-term financial markets.

Think back to our discussion of Amazon.com at the end of Chapter 2, "Fulfilling Orders: The Nature of Modern Order Cycle Management." Amazon is the world's most successful online retailer. Amazon's 2012 sales of $61 billion were greater than its next 12 largest rivals combined online sales (and that includes Walmart).[8] How does Amazon do it? Let's evaluate Amazon's approach using the *requisites* we noted above—vision, investment, and patience:

- **Vision**—From the beginning, Jeff Bezos was intent on building more than a retail empire. His goal was, and remains, to change the way people shop—and live. Amazon has never lost sight of this goal.

- **Investment**—Amazon has invested in a state-of-the-art Web interface that builds a personal "relationship" with individual customers, remembering when they last visited and what they like to buy. But, as noted in Chapter 2, the investment extends to a growing network of state-of-the-art fulfillment centers. In 2011, Amazon added 22 fulfillment centers worldwide (10 in North America, 12 overseas) totaling 15.2 million square feet. In 2012, Amazon opened another 9 fulfillment centers in North America and 10 internationally. Overall, Amazon operates 89 fulfillment centers across the world.[9] That is a lot of investment for a "virtual" retailer. However, with each new fulfillment center, the service target of same-day delivery is a step closer to reality.

- **Patience**—Emotional fortitude characterizes Amazon's CEO Jeff Bezos. His laser focus on the long-term vision enabled Bezos to stand against investor pressure to manage to short-term profitability. The early fortitude changed the way investors evaluate Amazon. Anthony Chukumba, equity analyst at BB&T Capital Market, explains, "If [Amazon] was a retailer, Jeff Bezos would have been fired years ago. There's a fundamental disconnect between how investors evaluate Amazon and how they evaluate brick-and-mortar stores."[10] In its battle for retail dominance, Amazon possesses a rare advantage: the ability to "run on thin or nonexistent profit margins without investors blinking an eye."[11]

The *hard* of Amazon's persistent investments in network configuration over almost two decades has made Amazon great. By contrast, rivals like Staples, Target, and Walmart have shown an on-again-off-again commitment to an understanding of what it takes to establish a customer-pleasing online fulfillment infrastructure. Target, for example, initially felt it could leave the *"hard"* to someone else, and chose to outsource its Internet channel to Amazon. Only in 2011 did Target take back control of its online destiny, relaunching Target.com. Target now highlights double-digit growth in online sales—but on an incredibly small base. When asked by the Securities and Exchange Commission to disclose actual sales numbers, Target responded, "digital sales represented an immaterial amount of total sales."[12] As for Walmart, despite a decade-long (and often repeated) commitment to winning on the Web, *The Wall Street Journal* summarized Walmart's Internet track record as follows:

> Walmart... *still hasn't figured out how to economically deliver all its products into the hands of online shoppers...* . It is a remarkable admission for the Bentonville, Ark., company, which became the world's largest retailer in part by the efficiency of its supply chain.[13]

The words "still hasn't figured [it] out" communicate the fact that configuring out an operating network for successful fulfillment is *hard*. What does this mean for you? The *hard* of order fulfillment can become a strategic weapon in your battle for profitable growth. If you can figure out what the right infrastructure looks like and then make the needed investments to turn concept into reality, the odds are good that you will achieve a competitive advantage. Few rivals possess the vision and the patience to go head-to-head and compete via outstanding order fulfillment. However, to build a supply chain network and order fulfillment infrastructure that enables you to meet customer expectations better than the competition, you need to grasp two vital facts:

- Infrastructure configuration is a complex, nontrivial task. Further, traditional organizational structures and cultures create performance tradeoffs and a resistance to change that work against the development of effective, efficient, and unique order fulfillment infrastructures.

- Therefore, you need to think differently than most decision makers, establishing a systematic approach to designing and developing a winning order fulfillment capability.

The Nature of Network Configuration

If outstanding order fulfillment were easy, everybody would do it. However, Walmart's so-far unsuccessful quest to conquer e-commerce shows that winning through order fulfillment is not easy.[14] Despite promising since 2001 to become a force online, Walmart racked up only $7.7 billion in 2012 online sales compared with $61 billion at rival Amazon.com.[15] It is important to remember that Walmart is no slouch when it comes to distribution. As early as 1992, the *Harvard Business Review* identified distribution as Walmart's "critical capability," noting that Walmart was pressuring rivals and capturing market share by delivering the right product to the shelf twice as often and at lower costs than its key rivals.[16] Nor is the Internet Walmart's first "new" order fulfillment system. In 1988, Walmart decided to get serious about selling groceries, opening its first Supercenter—a combination discounter and grocer. By 2001, Walmart had become the largest food retailer in America with grocery sales of over $56 billion.[17] Despite its track record of success, during the 2013 annual shareholder's meeting, CEO Mike Duke commented on Walmart's e-commerce challenge, "We're starting to gain traction. I say starting because we know that it's an area we *still have a long ways to go*."[18]

Complexity Is Everywhere

One reason that Walmart *still has a long ways to go* is that order fulfillment systems are incredibly complex. Consider the following facts about Walmart's existing (2013) order fulfillment infrastructure:

- Walmart manages more than 10,900 retail outlets under 69 banners in 27 countries.[19] In the United States alone, Walmart had 4,713 stores (3,211 Supercenters, 539 Discount Stores, 621 Sam's Clubs, 316 Neighborhood Markets, and 26 Walmart Express).[20] In all, Walmart managed over 721 million square feet of retail space—an area 10 percent larger than the entire city of Manhattan.[21]

- Walmart operates over 150 distribution centers in the United States. These DCs can be classified into six facility types:
  - General Merchandise Distribution Centers
  - Grocery and Perishable Food Distribution Centers
  - Fashion Distribution Centers
  - Import/Redistribution Centers (half of which are operated by 3PLs)
  - Specialty Distribution Centers (25% of which are operated by 3PLs)
  - Sam's Club Distribution Centers (over half are 3PLs)[22]

For perspective, consider that each of the 140 U.S. general merchandise DCs existing in mid-2013 supports 90–100 stores located within a 200-mile radius.[23] The average distance to each retail store is about 124 miles.[24] Each GMDC is more than one million square feet in size (approximately 18 football fields under one roof) and uses more than 5–12 miles of conveyor belts to move hundreds of thousands of cases through the facility every day.[25]

Of note, only about 80 percent of the merchandise sold through Walmart stores and 63 percent of the product sold through Sam's Club outlets passes through Walmart's DC network. The remainder moves through a third-party or via direct store delivery.[26]

Walmart's current supply chain network is very complex:

- Walmart's domestic fleet consists of 6,500 tractors, 55,000 trailers, and more than 7,000 drivers.[27]
- Walmart stocks an average of 142,000 distinct stock-keeping units (SKUs) at each Supercenter.[28] Add in store-by-store differences across 27 countries and Walmart's online options and Walmart probably manages a constantly changing quarter million SKUs.
- Walmart buys product from over 100,000 suppliers, which Walmart classifies as national product suppliers, local product suppliers, service and nonresale suppliers, and direct import and global suppliers.[29] Approximately 1,280 suppliers have set operations close to Walmart's headquarters in Bentonville, Arkansas, to provide better merchandising and logistics support.[30]

- Every week, 245 million customers make a purchase at a Walmart somewhere around the world.[31]

Even this very basic description of Walmart's operating network feels complex. Have you kept track of all of the different parts—let alone the specific numbers? Yet, that is what Walmart has to do every day. That is, to satisfy all of those customers—and keep them coming back week after week—every part of Walmart's network must work together smoothly and efficiently (much like auto parts in Honda's "The Cog" commercial). A breakdown at a supplier, in a DC, on the road (or sea), or within the store means that Every Day Low Price (EDLP) products will not be on the shelves when the customer wants to buy them.

As managers at Walmart—and almost all companies—have discovered, an essential step toward reducing the negative impact of complexity on order fulfillment is to understand the sources of complexity. Most companies must deal with seven sources of complexity:

- **Organizational structure**—The way a company is organized—for example, centralized or decentralized—affects complexity. Centralization leverages scale and enables more effective network rationalization (see Table 4-1).[32] Decentralization leverages localness to enhance customization and enable rapid response. Most companies today try to operate hybrid organizations to get the best of both worlds; however, this is not as easy as it sounds. Issues regarding decision-making authority, information availability, and measurement make coordinating hybrid structures difficult.

Table 4-1   Pros/Cons of Centralized and Decentralized Organizational Structures

Centralized Structure		Decentralized Structure	
Pros	**Cons**	**Pros**	**Cons**
Increases leverage	Increases bureaucracy	Has knowledge of local needs	Reduces leverage
Reduces duplication	Reduces flexibility	Has better local relationships	Leads to duplication
Promotes standardization	Can lose touch with reality of distinct operating environments	Has greater responsiveness	Is relatively inefficient
Enables specialization			
Provides greater control			

Adapted from: Fawcett, Ellram, and Ogden (2007)

For instance, e-commerce at Walmart is run out of its own headquarters in San Bruno, California. The goal was to locate closer to the critical hub of technological innovation. However, the distance—both geographic and cultural—has made it difficult for leaders to persuade Bentonville to make adequate investments in Internet distribution capacity. One former e-commerce executive notes, "plans were never executed and management would say the sales weren't there to justify the investment capital. Even now e-commerce is a rounding error in the U.S. market."[33] One result: Walmart's e-commerce operations were forced to rely on "makeshift spaces carved out of store-serving distribution centers," which increased costs to $5–$7 per parcel (67 percent to 75 percent higher than Amazon's per-parcel cost).[34]

- **Value-added processes**—By necessity, the order fulfillment process is complex and hard to manage. The "necessity" emerges from the fact that the overall order fulfillment process consists of many distinct activities and is cross-functional. For many companies, it is also global, which increases cycle times, decreases visibility, and adds another layer of complexity. As noted previously, the right product does not reach Walmart's shelf on time unless inventory, buying, warehousing, transportation, and technology decisions all work flawlessly together. Each "handoff" increases the opportunity to drop the value-add of products, driving satisfaction down and costs up. To mitigate this possibility, you must relentlessly streamline processes, invest in coordination mechanisms, and align performance measures.

- **The operating network**—Your operating network likely consists of a subset of the following: manufacturing facilities, distribution centers, and retail outlets. Your goal is to design a network that minimizes capital investment in infrastructure even as you pursue the dual operating goals of (1) satisfying customer needs at (2) the lowest possible day-to-day operating costs. You will want to consider four network-design questions:

    - How many facilities do you need?
    - Where should they be located?
    - What activities will be performed in each?
    - How will you coordinate the value-added activities?

    Answering these questions requires that you consider a variety of tradeoffs. For example, in Walmart's case, adding an additional DC dedicated to e-commerce meant committing to up-front investment in both bricks and mortar and technology systems. Until recently, decision makers in Bentonville have been hesitant to make this investment and have worried that low online sales volumes coupled with the opportunity to reduce delivery costs do not justify the investments. Very few companies are willing to accept unprofitable revenue growth! This is an advantage to Amazon. Another tradeoff Walmart is now considering is how

DC fulfillment compares with from-store fulfillment in terms of cost and service. The reality is that about two thirds of American households are located within five miles of a Walmart store.[35] Lacking the dedicated online DC network, can Walmart compensate—or create an advantage—using its existing retail network?

Assessing all of the tradeoffs typically requires intensive analysis to achieve an optimal network solution. Companies like Caterpillar and Whirlpool have used sophisticated mathematical modeling approaches like infinitesimal perturbation analysis to redesign their distribution networks.[36] At Whirlpool, the challenge was more difficult than normal as it involved the acquisition of Maytag and required the integration of the two firm's networks (i.e., 18 factories, 16 regional distribution centers, and 155 local distribution points). The redesign and integration shuttered or relocated a third of the premerger facilities. Senior management had promised Wall Street $120 million in three-year savings in freight and warehousing costs—a goal that the redesign exceeded.[37]

One final thought: You need to remember that optimality is a moving goal. Today's optimal solution becomes obsolete with tomorrow's technological disruption; a significant change in exchange rates, oil prices, or labor rates; the emergence of a new supplier; or a change in customer tastes/requirements.

- **Stock-keeping units**—Getting the number of SKUs right is critical. Too few, or the wrong items, and your company will lose sales and customers to rivals. Too many and your costs go up—both in manufacturing and logistics. Moreover, P&G discovered that too many SKUs confuse customers. P&G gave customers a sample product and asked them to find it in the store. Even though the product was on the shelf, only half of the customers could find it. This confusion goes beyond the decision paralysis that comes with excessive-SKU induced ambiguity.[38] P&G's conclusion: Reduce the number of SKUs in each product category.

Walmart has also faced vital SKU-related challenges in recent years:

- First, Walmart attempted to reduce SKUs and in-store clutter, eliminating 8,000 SKUs. However, Walmart faced serious customer backlash. Walmart discovered that many of its shoppers expect Walmart "to be everything to everyone" and reintroduced the SKUs to store shelves.[39]
- Second, "being everything to everyone" means offering EDLP pricing on a vast number of SKUs (142,000 per Supercenter). However, Aldi, the German deep discounter, has found a formula to undercut Walmart on price—by as much as 15 percent to 20 percent.[40] Aldi's secret? Aldi operates a small-format store (17,000 square feet), sells a limited number of SKUs per store (approximately 1,000) of which 95 percent are store brands, and minimizes labor costs by selling product straight from the packing carton and renting carts

to customers. Customers only get their money back if they return the cart to the rack. Walmart's response? Open Walmart Express—a 15,000 square foot, 15,000 SKU small format store.[41] Will it work? Time will tell. But, Express gives Walmart access to urban markets as well as to rural communities too small to support one of Walmart's traditional formats.

Getting SKU management right is difficult. As the following approaches demonstrate, careful analysis and discipline are critical:

- ***Rationalize SKUs***—Use activity-based costing to analyze cost-to-value contribution of every product, eliminating unprofitable SKUs.

- ***End SKU proliferation***—One company instituted a simple, but powerful policy: For every new product marketing wants to bring to market, it must identify one to be eliminated from the product portfolio.

- ***Employ a postponement strategy***—By postponing final manufacturing, assembly, packaging, labeling, or other value-added activities until a customer order is received, you can reduce SKUs dramatically. Consider Land's End's cotton slacks challenge. Table 4-2 shows that over 61,000 distinct SKUs could exist.[42] To reduce its SKUs, Land's End orders unhemmed slacks, cutting them to length at its Dodgeville, Wisconsin, DC once a customer order is received. By hemming and cuffing pants to order, Land's End can reduce the total number of SKUs to about 3,000.

Table 4-2   The Impact of Product Options on Total SKUs: Twill Slacks Example

Product Characteristic	Number of Options
Color: Black, Blue, Brown, Gray, Green, Purple, Tan, Yellow	8
Fabric property: No iron; wrinkle resistant	2
Front: Pleated; Plain; Comfort; Side elastic waist	4
Inseam: Regular; Tall; Big & Tall	3
Waist: 30, 31, 32, 33, 34, 35, 36, 38, 40, 42, 44, 46, 48, 50, 52, 54	16
Length: 27–38 inches	10
Bottom: Cuffed or Plain	2
Total SKUs	8×2×4×3×16×10×2=61,440
Simplified SKUs via postponement:	8×2×4×3×16×1×1=3,072
Simplified SKUs via data mining:	1,300

Adapted from: Fawcett, Ellram, and Ogden (2007)

- ***Use data analytics***—"Big Data" enables companies to make sense of expanding supplier, manufacturing, and customer databases. For example, Land's End collects and analyzes customer purchase data, enabling it to offer certain sizes like Big & Tall in selected colors and waist sizes, further reducing the number of SKUs it has to buy, inventory, and ship.

One final thought: Effective SKU management can change competitive dynamics for companies that recognize that rivals' strategies are failing to give customers what they really want. The classic example of competing on SKUs is found in the auto industry. Henry Ford dominated auto sales by selling only black Model Ts—a low-cost strategy that represents the ultimate in complexity reduction. However, when Alfred Sloan opted for more complexity via a variety of models in different colors, GM won the battle for customers' hearts—and money. The battle played out in reverse 50 years later. This time, Honda and Toyota gained cost and quality advantages by limiting the number of models—and the options available on each. Customers decided that low costs and reliability were more important than bells and whistles.

- **The supply base**—As most companies source a large percent of their COGS—typically 30 percent for service companies and 50 percent to 80 percent for manufacturers—building a strong supply base is critical to long-term success.[43] Supply managers typically describe their job as finding and developing the right suppliers. *Right* means that a supplier can deliver low-cost, high-quality goods and services on time every time. Right also means the supplier is flexible enough to adapt to rapid market changes (including unforeseen disasters) and can provide needed technical support to help your company develop new products and services. Although this job description is demanding, the real challenge is to do this for every SKU your company buys. That means managing hundreds to thousands of supplier relationships (in Walmart's case, it's over 100,000). Every transaction must be managed efficiently. Every relationship must be appropriately defined and developed: Many will be short term and cost focused; others must be long term, trust-based, and collaborative.

  Because resources are constrained and the success of many of your company's strategic initiatives depends on the strength of your supply base, you need to take a strategic approach to supplier segmentation.[44] For most companies, the desire to free up resources to work more collaboratively with the most important suppliers has led to an effort to rationalize the supply base. Xerox set the standard in supply-base reduction. To stem market share losses to rival Canon, Xerox determined that it needed help from its supply base in order to bring lower-cost, higher-quality photocopiers to the market—and it needed to do so with shorter research and development (R&D) times. However, trying to collaborate with 5,000 global suppliers was an impossible task. Xerox examined its spend, consolidated purchase requirements—which increased its negotiating leverage with suppliers—eliminated redundant suppliers, and invited its most important suppliers to become partners in value creation. The result: Xerox slashed its supply base to just over 400 suppliers, reducing direct expenditures as well as overhead rates (from 9 percent of total materials cost to about 3 percent).[45] Now, a warning: The takeaway for many outside observers was that they too should strive to reduce their supply base by 90 percent. This arbitrary 90 percent became a frequently targeted goal. The more appropriate goal is to get the right number of the right suppliers on a category-by-category basis—even if that means adding to the existing number of suppliers.[46]

Walmart adds a couple of twists to its approach to managing complexity in the supply base. Like other companies, Walmart works closely and collaboratively with select suppliers on major initiatives. For example, it worked closely with 100 top suppliers to test the viability of RFID.[47] However, Walmart has also experimented with novel organizational structures to spur innovation among its supply chain network. For instance, Walmart set up 14 Sustainable Value Networks (SVNs) consisting of employees, suppliers, government agencies, academics, and Non-Governmental Organizations (NGOs) to create a collective center of knowledge and a reservoir of passion to help Walmart meet aggressive sustainability goals.[48] Two other practices merit brief mention:

- *Category captains*—With so many SKUs to manage, Walmart (like many retailers) has outsourced merchandising decisions for entire categories of SKUs to "category captains." Category captains use deep expertise that comes from industry-specific experience and extensive consumer data to provide Walmart recommendations on what products to carry, how and where to display them, and even what pricing should be. Effective use of category captains can increase category profitability substantially. However, the use of category captains is not without risk. A category captain could be tempted to provide subtle advantages such as a slightly better shelf placement for its products. Some retailers have employed secondary captains to evaluate the quality and objectivity of the primary captains' recommendations.[49]

- *Colocation of suppliers*—Over the past two decades, a number (approximately 1,280) of Walmart's most important suppliers have established offices in northwest Arkansas so that they can better support Walmart's merchandizing and logistics needs. This colocation (which is sometimes nicknamed "vendorville") has created a collaborative learning cluster that speeds innovation across the entire Walmart and supplier community.[50]

- **The customer base**—As challenging as supply-base rationalization is, your company probably has a more diverse and complex customer base. However, your company culture might not seek to limit sales to some customers, based on their profitability. It is tough to walk away from a customer's business. Rather, the focus (especially if the sales force works on commission) tends to be on growing sales. Moreover, the power dynamic is different—customers possess much more power than the typical supplier. If your efforts to manage the customer base create dissatisfaction, the customer can always take her money to one of your rivals. This detail makes managing customer relations extremely tricky. As a rule, customers do not like to be treated the way they treat their suppliers. Even so, if you think back to our discussion in Chapter 3, "Developing a Winning Customer Fulfillment Strategy," you understand why customer segmentation and profitability analysis are critical to the design of an effective order fulfillment and customer service infrastructure. The most common approach to customer rationalization to date focuses on the use of loyalty cards and customer relationship management systems. We will discuss this in greater detail in Chapter 5, "Implementing an Enabling Technology Strategy."

As we have used Walmart to exemplify approaches to managing the sources of complexity, we should note that Walmart has yet to adopt a loyalty card. When asked why Walmart does not have a loyalty program, Charles Holley, executive vice president and CFO, replied, "We believe that all of our customers deserve the lowest price possible, not just certain customers."[51] However, Walmart is testing a mobile phone app called "Scan & Go." The app allows customers to scan each item with their phone as they place the item in their cart. At checkout, customers sync their phone with the register and pay electronically. If the app catches on, Walmart will be able to collect even more data than is available via a loyalty card, including how much time customers spend shopping and the route they take through the store. Further, getting customers to contribute their own labor to the checkout process could lead to big cost savings. Holley noted, "For every second we can save in Walmart U.S., that equates to $12 million in savings."[52]

- **The logistics system**—Your company's logistics system ties the entire network—suppliers, production facilities, warehouses, and customers—together. The number of options among modes, carriers, and routes is a combinatorial nightmare. Add in variations in product demand and customer-specified delivery time windows as well as the desire to minimize both carbon footprint and expensive empty backhauls. Finally, beyond the quest for efficiency, the order fulfillment logistics system is the final link in the customer service supply chain. No wonder SC managers describe their logistics systems as a "tangled web."[53]

How are companies detangling the web? Two approaches dominate most companies' efforts:

- *Outsourcing*—Many companies have decided that the headaches induced by logistical complexity are not worth dealing with. They reason that they are in business to make a product or deliver a service, not to run a logistics company.

As a result, they turn to a third-party logistics expert. For instance, Sony helped pioneer creative value-added outsourcing. When it was considering how best to fix customers' broken laptops, it turned to UPS. When a customer calls, UPS' familiar brown truck stops by to pick up the broken laptop, which is shipped to the nearest UPS repair center. The problem is diagnosed, the laptop repaired or replaced, and the working machine returned to the customer the next day.[54] More recently, when GM needed better spare parts availability at lower costs, it called on Schneider National Logistics to redesign and then operate its spare parts distribution system.

- *Technology*—Today's sophisticated enterprise resource planning (ERP) systems include logistical planning modules that rely on mathematical optimization algorithms to define optimal solutions in terms of carrier selection, route planning, and vehicle loading. When combined with satellite tracking technology and radio-frequency tagging, companies can track product and equipment in real time as it moves through the network—from supplier to customer. Managers can make needed logistical adjustments as things as diverse as customer needs or the weather change.

To summarize, one of your most daunting tasks in building a reliable and flexible order fulfillment and customer service capability is to understand and mitigate the negative consequences of complexity. Complexity is everywhere and the confusion it creates increases costs and diminishes service levels. Worse, as economics columnist Megan McArdle has noted, "People tend to radically underestimate the costs imposed by complexity, because the management problems do not simply add up; they multiply."[55] However, you need to remember that not all complexity is bad. Remember Henry Ford's black Model T. Too late, Ford discovered that some additional complexity—even at additional cost—was needed to stave off an up-and-coming GM. What should you take away from this discussion? If you understand both your customers and how the sources of complexity affect your ability to meet their needs, you will be better able to manage the tradeoffs that pervade network design.

Tradeoffs Are Everywhere

A second reason Walmart has taken so long to build out its e-commerce order fulfillment infrastructure is that every decision creates numerous, often hidden, ripple effects that threaten to affect the rest of its distribution network. As Walmart has repeatedly discovered, tradeoffs are everywhere—and they are often hard to identify and to quantify.

For example, Walmart's latest efforts (2013) involve trying to leverage its 4,000+ stores to fill some online orders directly from local Supercenters.[56] Remember, two thirds of U.S. consumers live within a few-minute drive of a Supercenter.[57] Walmart thus believes it can deliver store-to-door at lower costs and with shorter delivery times than by copying Amazon's approach of relying solely on fulfillment DCs. But this strategy raises a serious question, "How will 'from-store' fulfillment affect in-store operations?" Without the distraction of running an e-commerce solution from their stores, most retailers struggle to keep their shelves stocked. Can workers really be expected to roll carts down aisles, pick online orders, and mail them out to customers while still keeping shelves stocked and serving customers—at a cost that competes with Amazon? So far, no major store chain has figured out this delicate balancing act.

Like Walmart, for you to design an order fulfillment system that wins orders and retains customers, you will need to examine a myriad of tradeoffs. What are the implications of each design decision on costs, lead time, dependability, and other performance dimensions? Although this sounds commonsensical, identifying and assessing the tradeoffs is no trivial task. The devil is in the details. This point should be clear from our previous discussion of complexity. Alone, each source of complexity presents difficult tradeoffs. When combined, the seven sources of complexity confound one another, making it difficult to identify all of the relevant tradeoffs—let alone analyze them.

Moreover, the way most companies are structured—to build and leverage deep skills—complicates your efforts to evaluate the tradeoff. Organizational boundaries make tradeoffs harder to manage proactively. Peter Senge, author of *The Fifth Discipline*, describes how structural distance and temporal delays insulate decision makers from the real consequences of their own decisions.[58] When the downside happens somewhere else (e.g., a different department or division), managers may never know their decisions raised a colleague's or a supply chain partner's costs. Boundary-induced diminished empathy can also lead a manager who initiates a sequence of counterproductive effects to view these costs as "somebody else's problem." The bottom line: Managers not only fail to understand how their decisions affect overall system performance, but they may also end up blaming someone else for the system disruptions that they actually caused. If you have played the Beer Game, you have likely experienced this phenomenon. Senge argues that in order to make good decisions, managers need to understand how structure, behavior, and events influence each other (see Figure 4-1).[59]

Considering structure, Figure 4-2 shows how classic goals and measures promote conflict and performance tradeoffs—both of which make it difficult to design cross-functional processes. The four functions that most influence the physical flow of the order fulfillment process are depicted: purchasing, production, logistics, and marketing. Marketing, for example, is managed as a profit center. Marketing managers are graded on their ability to increase sales. Not surprisingly, they promise customers exceptional order fulfillment and then demand/expect the rest of the organization/supply chain to deliver. Great order fulfillment often means short lead times supported by ample inventory staged as close as possible to customers. Thus, Amazon and Walmart have built expansive fulfillment networks. By contrast, logistics is a cost center. Logisticians have a mandate to drive costs down and are evaluated on their results. In an ideal world, this means long lead times supported by as little inventory as possible—held in as few centralized DCs as possible. Similar conflicts are found among all four functions creating a tug-of-war within the company as managers from each function pull toward short-term, local goals.

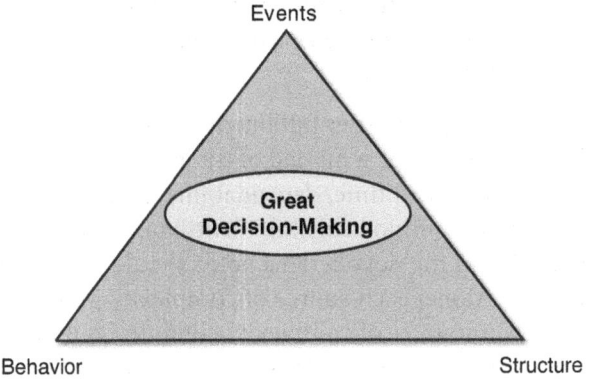

Figure 4-1   Drivers of great decision making

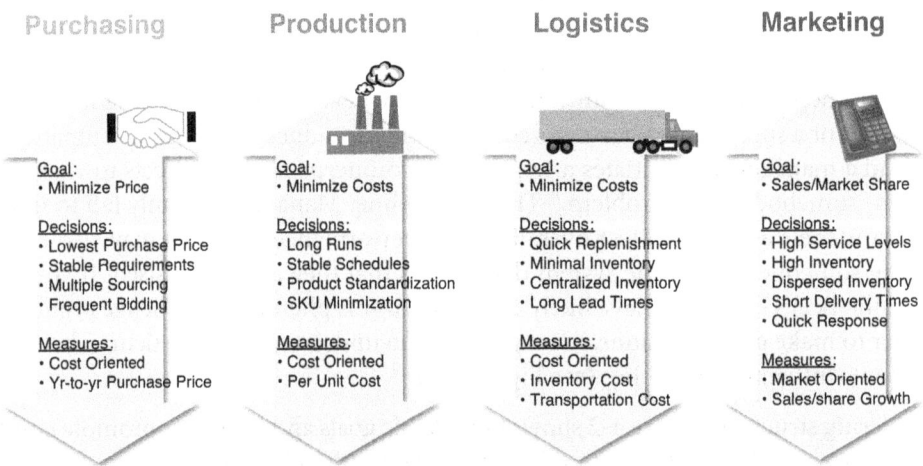

Figure 4-2   Structure's influence on functional decision making

Unfortunately, potential decision-making conflict is ubiquitous in order fulfillment and customer service. When UPS made the strategic move into outsourced fulfillment services, it used the graphic in Figure 4-3 to help its managers understand tradeoffs are inherent across all the decisions involved in order fulfillment. As you evaluate the

110   THE DEFINITIVE GUIDE TO ORDER FULFILLMENT AND CUSTOMER SERVICE

different decision areas, note that every decision leads to at least one negative consequence when compared with the global ideal shown in the bottom row. Your challenge is to weigh the tradeoffs and design a fulfillment infrastructure that achieves corporate strategic goals with the least disruption for individual departments across the firm.

Functional Objectives	Impact of objectives on . . .		
	Inventory	Customer Service	Total Costs
High customer service	↑	↑	↑
Low transportation cost	↑	↓	↓
Low warehousing costs	↓	↓	↓
Reduce inventories	↓	↓	↓
Fast deliveries	↑	↑	↑
Reduced labor costs	↑	↓	↓
Desired Results	↓	↑	↓

Figure 4-3 Tradeoffs inherent in order fulfillment and customer service

Focusing on behavior, consider how functional thinking is built into modern management. Think about the career path of the typical manager. Early in her college career, she is exposed to the various business disciplines, including finance, human resources, marketing, and operations. Then she chooses a major. Because she knows companies want to hire people with deep skills and she wants to find a job upon graduation, she spends the rest of her time taking classes in her chosen specific discipline. (Even if she wanted a more holistic education, program requirements make taking cross-functional coursework almost impossible. B-schools cater to recruiter requests for deep skills and monopolize students' "electives" with functionally oriented classes.) Throughout her coursework, she hears over and over that she made a good career choice because her discipline is the most important (e.g., finance is the language of business; marketing is the gateway to the customer; people are the only source of creativity). Once she lands her dream job, her employer continues her functional indoctrination. As specialization is vital to competitive success, her company needs her to possess deep skills. This quest for deep skills would be OK except for the fact that it leads her to forget that managers in other areas of the firm also create needed value. As Senge warns, "She becomes her function."[60] This behavioral tendency represents another tradeoff you will face. Mike Wells, former VP of logistics at Hershey explained, "If you ask me what I stay awake at night thinking about, its cross-functional processes. The challenge is to become more process focused while maintaining functional expertise."[61] Tradeoffs like this make an already complex task even more difficult.

Change Is Constant, but Hard

A third reason Walmart has struggled to establish a dominant online order fulfillment capability is that its bricks-and-mortar operations are fantastically successful. Walmart's 2012 operations produced $467 billion in sales, $28 billion in operating income, and over $17 billion in net income.[62] No retailer has ever come close to this level of market hegemony. It would take *eight* Amazons to match Walmart's global sales (even more to reach Walmart's net income).

As a result, despite the Internet's cool factor, Walmart leadership has treated it more as a plaything (occasionally a distraction) than a strategic focus. Consider the numbers. Online sales in the United States hit $224.3 billion in 2012[63] and accounted for 5.8 percent of overall retail sales in the second quarter of 2013.[64] Walmart's $7.7 billion in 2012 online sales are less than 1.7 percent of its total global sales. With all of the other issues confronting top management—for example, sustainability, recession, new market entry, corporate image, and continuing to run a successful bricks-and-mortar operation—it is easy to see why Walmart has never fully committed to the Internet.

Further, e-commerce represents an entirely new way of doing business for all traditional retailers—one that could eventually supplant business as usual. Thus, it is not surprising that managers treat it "as a rounding error" or that digital sales remain an "immaterial amount of total sales." One of Amazon's strengths is that it was never built to compete in the physical world of retail. It grew up as the paragon of virtual retailing, organized from the beginning as an online retailer. For Walmart, Target, and other bricks-and-mortar retailers, organizing for online competition requires a new business model, which means real change. In this respect, the challenge of establishing an e-commerce infrastructure is the *challenge of change*.

Figure 4-4 depicts force field analysis, which can help you understand why retailers have persisted in a steady, potentially counterproductive state—that is, a relentless focus on bricks-and-mortar operations in the face of an emerging online marketplace. Three common responses—and outcomes—to the need for change exist. Resistance and failure to adapt is shown in Panel A. When the world changes, managers sense an emerging opportunity or threat, feel a need to change, and pursue a new strategy (e.g., the development of an online fulfillment infrastructure).

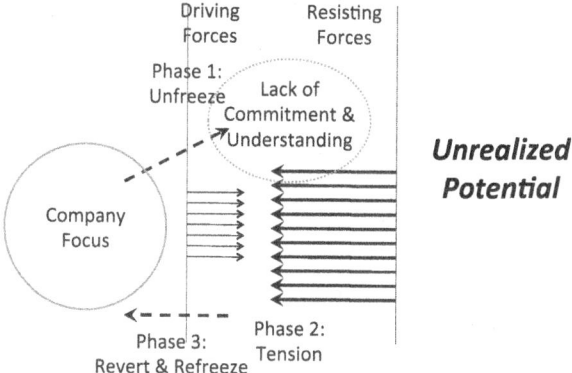
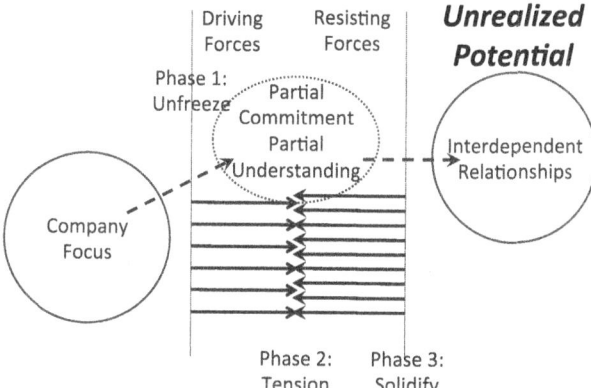
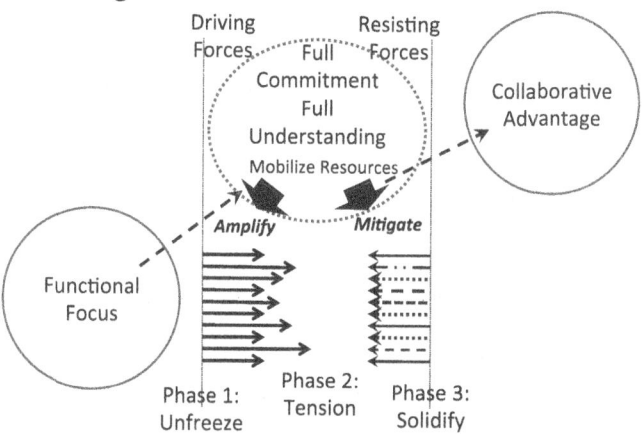

Figure 4-4 Understanding the dynamics of change

Chapter 4 Configuring the Network for Successful Fulfillment 113

The changing marketplace drives the firm into a period of potential transition phase in which new capabilities are sought. But, as soon as the firm begins to move into this transition phase, resisting forces (e.g., skeptics or competing resource demands) emerge and seek to prevent meaningful change. When resistors are stronger than drivers, change stalls and desired capabilities do not emerge. If you do not grasp the dynamics of change management, you may fail to mobilize the resources required to mitigate resisting forces (something that evidence suggests has befallen most bricks-and-mortar retailers). Repeated missteps may provoke cynicism, making it difficult to build buy-in for future change efforts. After a decade of false starts, it may be a little more difficult for Mike Duke to convince managers, suppliers, customers, and stock analysts that Walmart truly is committed to an online platform.

Panel B begins in the same initial state and pursues the same dynamics into the transition phase. Higher levels of managerial commitment to the change initiative (i.e., the Internet platform) and a better understanding of the change process helps managers commit a better mix of resources to attenuate the resisting forces. Progress is made toward the goal and benefits begin to emerge. With a forecast of $10 billion in online sales for 2013, you might argue that Panel B better describes Walmart's scenario. Even so, the limited levels of commitment as signaled by investment have left the digital vision unfulfilled. Tremendous unrealized potential persists.

Panel C communicates the required conditions for successful change: full commitment and full understanding. External driving forces are fully leveraged to disrupt inertia, making change possible. At the same time, adequate investment enables infrastructure development and changed behavior. In our case, external drivers come in two forms. First, hope for gain. Total Internet sales are expected to double by 2017. Second, fear of loss. Amazon's strategy of same-day delivery appears to be within striking distance of reality. This would be a real game changer for retail and an advantage that Walmart does not want to cede to Amazon. You may be interested to know that Harvard's Michael Porter has claimed that the fear of loss is a greater motivator than the hope for gain. Together, these two inflection points (rate of sales growth and level of rival's capability) provide ample motivation to take online sales more seriously. As for investment, Walmart leaders are speaking with more urgency. CEO Mike Duke has said that "winning in e-commerce" is one of Walmart's top six priorities.[65] Neil Ashe, Walmart's president of global e-commerce, has pronounced, "This is an opportunity for reinvention. We have some of the most talented logisticians in the world and we are playing to win."[66] When asked "how long it would take and how much it would cost" to build its e-commerce fulfillment infrastructure, Ashe responded, "It will take the rest of our careers and as much as we've got...This isn't a project. It's about the future of the company."[67] Walmart has thus announced that it will allocate approximately $430 million to infrastructure in 2013.[68] Dedicated DC space as well as in-store investments are planned. Pilot tests of store-enabled fulfillment are taking place at 50 stores.[69] Perhaps Walmart has discovered the right mix of motivation, commitment, and understanding to establish a fulfillment capability able to take on Amazon and win. Only time will tell.

To summarize, establishing order fulfillment and customer service infrastructures capable of delivering *WOW* experiences is hard. If it weren't, everybody would do it. You would consistently be delighted with the service you receive and the customer service index would be in the high 90s instead of the mid 70s. But, complexity confounds configuration decisions. Pervasive tradeoffs require better visibility to understand which compromises are needed. And making the radical—or even the consistent incremental—improvements needed to offer world-class fulfillment requires constant change. All of these moving parts make "perfect" orders and 100 percent on-time delivery BHAG (big hairy audacious goals), but unrealistic goals. To move toward these goals, you need to (1) rethink how the pieces fit together and (2) find a better way to weigh the consequences of infrastructure decisions. Better insight into interrelationships and decision outcomes will help you move forward with confidence. That is the essence and goal of systems thinking.

Systems Thinking and Order Fulfillment Configuration

Systems thinking—the ability to perceive interrelationships and discover structural drivers of behavior and results—can help you work through the order fulfillment design process.[70] Systems thinking warns that, "By separately optimizing individual activities, overall system performance is often sub-optimized."[71] This is the core message from Figures 4-2 and 4-3. Systems thinking helps you avoid suboptimization by helping you take a holistic approach to decision making. Systems thinking requires that you define the system, identify the relevant parts, and figure out how they work together. Only then will you grasp the systemwide ramifications of individual technology implementation, infrastructure investment, and other process design decisions. Further, to avoid the negative effects of the law of unintended consequences, systems thinking emphasizes how important it is to consider longer-term effects. That is, systems thinking explicitly emphasizes that decision making is a chain reaction. As in tumbling dominoes, you need to carefully evaluate how each domino will fall if you want the sequence to conclude successfully.

The following six steps for analyzing and managing a system can help you apply systems thinking to order fulfillment:

1. **Establish the superordinate goal**—Tom Peters, author of *In Search of Excellence*, stressed, "What gets measured, gets done."[72] He is right. People work toward expressed goals that are measured. Consider a well-documented fact: Organizational conflict emerges "when one party perceives that the other has negatively affected, or is about to negatively affect, something that he or she cares about."[73] Such decisions—that "negatively affect" other members of a value system—occur most frequently when local goals are competing rather than complementary.

Because goals guide behavior across your order fulfillment and customer service systems, you need to carefully choose the right superordinate goal. You do this by rigorously defining desired system outcomes. Confidence in what the entire team is trying to accomplish allows you to set and communicate the superordinate, systemwide goal. All subsystem goals can then be aligned and appropriate measures put in place. Complementary goals that are well understood (a critical role of effective measurement) help decision makers reconcile the tradeoffs they are bound to encounter. Remember, tradeoffs are everywhere, beginning with the fundamental operational objectives of fulfillment and service systems—to maximize service and to minimize costs. The bottom line: What people perceive as your real goal will influence every aspect of your system, including the decisions that are made and the outcomes that result.

2. **Define system boundaries**—What you are trying to accomplish defines your system's boundaries. Systems can be defined at any level—from an in-warehouse picking system to a global supply chain. System boundaries determine who is on the value-creation team. For example, the SCOR delivery process D1 (introduced in Figure 2-4) denotes that the order fulfillment system includes different departments within your firm (e.g., order management, transportation management, warehousing) as well as customers and logistics service providers. To achieve delivery goals, each member of the system must perform specific roles and responsibilities, which are defined in the next four steps.

3. **Determine elements and interrelationships**—Once you define the order fulfillment system's boundaries, you are in position to identify the specific activities/elements that make up the system. Value-stream mapping philosophies and tools can be very helpful here. The reality is that for many firms, the specific activities/steps in key processes are unknown. No one has meticulously mapped the process to create the required visibility. After the activities/elements are identified, you can begin to determine how they interrelate. Your challenge is to figure out how a decision made in one area affects decisions and operations in the other areas of the fulfillment system. Be sure to look across the entire system to identify systemwide implications. Again, process visibility supported by data analytics can help you assess the numerous, diverse interactions in a complex order fulfillment system like those managed by Amazon and Walmart.

4. **Gather and delimit information**—As the previous steps imply, systems thinking is information intensive. Specifically, information regarding customer needs, the competitive setting, and organizational capabilities defines goals and systems boundaries. In the old days, collecting data was a serious—and often overwhelming—impediment to effective systems analysis. Advances in data science—that is,

Big Data, data analytics, and advanced sensor technologies—allow you to build incredibly rich and detailed databases—all but eliminating the data insufficiency dilemma. Today, your challenge is to avoid data overload as you seek to understand the intricacies of system behavior and performance. To do this, you need to:

- Decide exactly what information you need.

- Determine how it can best be captured and analyzed.

- Identify who needs the information and how to get it to them in a timely manner.

5. **Evaluate tradeoffs**—By nature, complex systems are characterized by tradeoffs. Few decisions are made in isolation. Decisions induce side effects across the system. To make a sale, marketing may promise delivery without verifying product availability. To make good on the promise and avoid alienating a valuable customer, expensive overtime in manufacturing or expediting in logistics may be required. When this happens once, the profitability of a sale may be diminished. When such behavior goes uncorrected and becomes habit, chaos in the fulfillment system undermines the firm's delivery capability and competitiveness. Of course, the most obvious tradeoffs are typically recognized and avoided; however, numerous tradeoffs go unseen. Nobody really knows whether current decisions—or existing infrastructure—are appropriate. Your challenge is to make relevant tradeoffs visible and open for inspection via total costing so that the brutal facts can be confronted and blameless autopsies conducted. Absent such analysis, you will consistently suboptimize systemwide performance.

6. **Take constraints into account**—No value-added system operates without constraints. Constraints limit options and exacerbate tradeoffs. Perhaps the most common constraint is a lack of resources. Mike Duke has identified e-commerce fulfillment as one of six strategic priorities. That means that managers responsible for establishing the e-commerce infrastructure must compete for investment with managers from at least five other strategic domains. The resulting physical constraints can dramatically reduce operating efficiency and drive costs up and service down. Sometimes companies impose constraints on their systems via unnecessary policies and procedures. Poor habits or counterproductive behavior also constrain systems. You should also look beyond your firm to identify relevant constraints. For example, customer requirements, supplier capabilities, and government regulations can hinder system performance. Because diagnosis precedes prescription, your task is to identify root-cause constraints and diagnose their influence on system behavior and performance. You will then be able to suggest steps to alleviate them.

Properly pursued, systems thinking can help you build unique order fulfillment and customer service solutions. Two routes are exemplified in the following list:

- **Competitive strategy**—For Sony de Mexico (see Chapter 2's opening story), systems thinking helped the company avoid closure by refocusing the company's strategic orientation. Born out of a desire to minimize costs, the managerial mind-set had always focused on keeping costs down. When competitive dynamics changed and the cost advantage transitioned to Asia, managers at Sony de Mexico could no longer justify their existence. All of the cost arguments fell flat. However, as soon as the system goal was reframed to resolve customer pain points, new options emerged. By designing shrinking fulfillment cycles, Sony de Mexico began to compete on speed—something Asian operations could not match. Managers had used a systems-thinking approach to identify a reason to exist.

- **Operational efficiency**—For National Semiconductor, systems thinking helped overhaul an inefficient global fulfillment system. The redesign began when managers realized, "We had no idea what this process [fulfillment] was costing us."[74] Managers traced product as it flowed through National's geographically dispersed network, which included six fabrication plants, seven assembly operations, and six warehouses. Activity-based costing was employed to evaluate the profitability of each SKU in its product line. What did they learn? Delivery cycles were hugely variable, the system was awash in inventory, and, amazingly, half of National's product line "didn't generate a damn in revenues or profits."[75] What did National do? It eliminated 45 percent of its SKUs and replaced six regional warehouses with a single, 3PL-run DC in Singapore. The results: 47 percent shorter delivery times helped National increase sales by 34 percent—all at 2.5 percent lower distribution costs.

Global Implications for Network Configuration

Walmart ventured into international markets in 1992, establishing operations in Mexico and Canada. Company leadership felt confident, saying, "We believe all of our principles and many of our concepts are exportable."[76] Three years of losses followed—but Walmart's business model carried international operations to a $24 million profit in 1995. Confidence intact, Walmart expanded into Argentina and Brazil. Now, however, Walmart has had to build its fulfillment infrastructure from scratch. Equally daunting, an experienced rival, Carrefour, greeted Walmart with surprising tactics. Tomas Gallegos, a Walmart store manager in Buenos Aires, was startled when he printed up fliers advertising bargains only to see the nearby Carrefour pass out its own fliers undercutting his prices on the same products within a couple of hours. Worse, Carrefour handed out its fliers at the entrance to Walmart's parking lot. Gallegos' reaction: "Geez, is the competition aggressive."[77]

Walmart hurt its own cause by making some silly merchandizing mistakes: Brazilians play futbol not football and leaf blowers are "useless in a concrete jungle such as Sao Paulo."[78] More critically, Walmart underestimated the challenge of doing business without its vaunted business model! Lacking scale, Walmart couldn't justify investing in its own trucking fleet. Forced to rely on suppliers and third-party carriers, Walmart lost control over deliveries and on-shelf performance—a key element of its hyperefficient, customer-pleasing fulfillment system. The lack of scale also hampered Walmart's ability to build a strong supply team. Domestic purveyors of uniquely local products like dulce de leche, a must-have for Argentine shoppers, were not eager to supply Walmart. At one point, 11 local suppliers, who viewed the Walmart way as the wrong way, refused to sell to Walmart. Even South American affiliates of Walmart's U.S. suppliers were reluctant partners. The close relationships established in the United States were not instantly exportable. David Glass, Walmart's CEO, explained the pain, "It's slow going early on, and you spend a lot of money. You pay a lot of tuition to learn what you need to learn."[79]

What did Walmart learn? One point stands out: Walmart learned that "all of its principles" are *not* exportable. Doing business globally is different! For instance, culture matters, influencing expectations across customers, suppliers, and employees. Infrastructure is also different, affecting operating efficiency and performance. These facts were obscured by Walmart's early approach to internationalization. Walmart bought its way into the Canadian market, acquiring 120 Woolco discount stores. To gain a foothold in Mexico, Walmart partnered with Cifra, a well-established Mexican retailer. Experienced partners possess the inside skinny, key relationships, and infrastructure to ease the pain of adapting to global markets. However, when Walmart attempted to go it alone, unanticipated and unplanned for differences raised adaptation (i.e., tuition) costs. The need to recognize and adapt to key operating differences was driven painfully home by embarrassing failures in the German and Korean markets.

Doing your homework to know when and what to adapt can help you design a winning global fulfillment infrastructure. Taking a systematic approach helps make sure you get your homework right. Five decision areas—compatibility, configuration, coordination, control, and continuity—should be assessed to help you design a more competitive global network.

Compatibility

Compatibility is all about strategic fit. Does it really make sense to enter a specific market? For example, when Mercedes and BMW decided to build production capacity in North America, Mexico was clearly the low-cost option. Yet, decision makers decided that "Hecho in Mexico" was the wrong label for German luxury cars. Sufficient cost savings were found in Alabama and South Carolina, respectively. Many companies, however, do not do an adequate job of assessing compatibility. They are too preoccupied with chasing ever-lower wage rates or following the competition. If you don't carefully assess strategic compatibility, your entry costs will definitely go up (e.g., think Disney's choice to build Euro Disney in France instead of Spain). Similarly, your probability of success will go down (e.g., think Walmart's entry into Germany). Although it seems like common sense, you need to carefully assess every new network decision to ensure that it supports your proposition.

Configuration

Configuration decisions determine *where* value-added activities are performed. The classic decision is facility location. In the preceding Mercedes example, Mercedes considered several possible production sites, looking for an abundant, hard-working, and nonunionized workforce. Mercedes also sought access to high-quality suppliers and a well-developed transportation infrastructure. Issues like the cost of real estate, utility rates, and tax incentives also played into Mercedes' final decision to build in Alabama.

Configuration also involves decisions regarding the allocation of value-added activities to specific locations. For example, to reduce the estimated $10 billion up-front cost of developing the 787, Boeing invited supply partners to take ownership (i.e., invest their own money) in the design and manufacture of major portions of the plane. Vought Aircraft Industries produced the rear fuselage at a facility in South Carolina. Alenia Aeronautica manufactured the middle fuselage and horizontal stabilizers in Italy. Japan's Mitsubishi designed the 787's stylishly swooped wing. The idea was for each subsystem to arrive in Everett, Washington, ready to be snapped together like pieces of a model airplane. Unfortunately, in the words of Boeing CEO Jim McNerney, "The initial plan outran our ability to execute it."[80] The first plane came to the facility in 30,000 pieces.[81] Rather than a snap-together model, Boeing had a monster-sized jigsaw puzzle. Boeing eventually acquired Vought's South Carolina manufacturing facility to alleviate late fuselage deliveries. Far from reducing up-front costs from a hefty $10 billion, poor allocation of responsibilities ballooned costs to an estimated $32 billion—half for development and half for manufacturing.[82] So focused on reducing up-front costs, Boeing failed to do due diligence in designing its 787 business model.

Coordination

Coordination focuses on how to connect geographically dispersed activities into a cohesive value-added network. The goal is to leverage the strengths of each location or partner without driving up costs excessively. You need to remember, however, that basic coordinating mechanisms such as information sharing and transportation are almost always more difficult to manage in the global arena than in domestic operations. Physical distance, poor infrastructure, cultural uniqueness, and bureaucracy create confusion; raising costs and lowering performance. For example, Texas Instruments (TI) brought in a consultant to evaluate its global network. The findings were twofold. First, TI's global configuration was vital to competitiveness. Second, poor coordination resulted in excessive backtracking, poor equipment utilization, and high levels of expediting—increasing logistics costs by over $100 million per year. Both TI's and Boeing's experiences highlight the need to evaluate configuration and coordination issues concurrently.

Control

Control involves managing day-to-day operations for maximum value creation. Global operations can improve top- and bottom-line performance by providing access to customer and resource markets. However, they only improve competitiveness if decisions made onsite—at production facilities, distribution centers, and supplier operations—lead to better productivity, quality, and service. Yet, as previously noted, the devil is in the details—and the differences. Differences in language, legal issues, reward systems, workforce relations, and infrastructure complicate day-to-day operations very quickly. The depth of detail required to move product around the world in a timely and efficient way leads many companies to rely on various third parties such as freight forwarders, customs house brokers, and logistics service providers. Likewise, the specialized know-how and connections needed to maintain good relations with global stakeholders has convinced many firms to employ country-specific consultants.

Continuity

Continuity is all about keeping operations running when things go wrong. The fact is that efforts to run lean global networks have placed many supply chains at risk. This reality became widely recognized when a fire burned down an Aisin brake plant in 1997, shutting down Toyota assembly lines within 4 hours of the disaster. Since then, a wide variety of events have caused supply chain disruptions (see Figure 4-5). You may find the following facts regarding supply chain risk startling:

- A large majority of surveyed companies (70 percent to 99 percent depending on the study) report that they have suffered a supply chain disruption during the previous year.[83]

- Most surveyed managers (48 percent to 73 percent depending on the study) indicate that risk levels have increased in recent years.[84]

- More than half (53 percent) of surveyed executives report that supply chain disruptions are more costly than previously.[85]

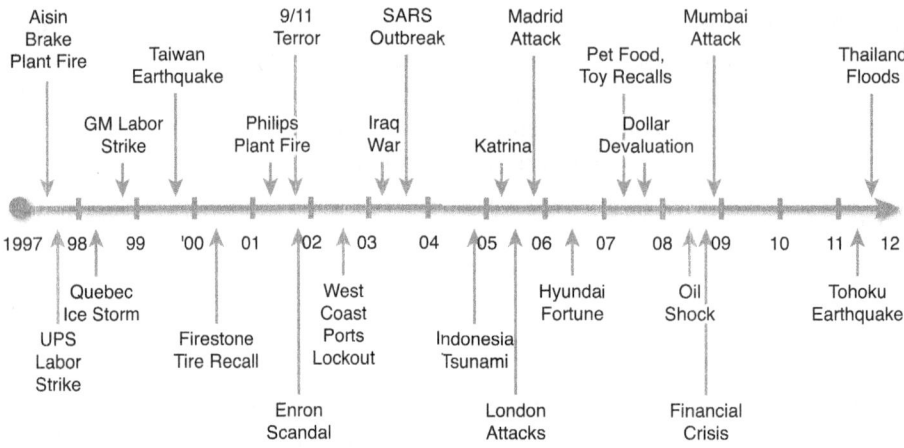

Figure 4-5   Recent causes of supply chain disruptions

Perhaps most startling, despite the heightened awareness of the threat and cost of supply chain disruptions, few managers feel their companies have highly effective continuity plans in place.[86]

As with the previous design issues, continuity planning is information intensive. It begins with a risk assessment that identifies potential risks and then assesses their potential threat in terms of probability and impact (see Figure 4-6). To be truly effective, this assessment must be done for each manufacturing plant, DC, and A-level supplier and customer in your network. Once the risks are identified, plans must be made to cover all of the likely eventualities. Now, you know why so few executives feel their organizations have well-thought-out-and-articulated continuity plans in place.

Now, a little good news: If doing your homework enables you to establish a global network that can deliver both the goods and high levels of customer satisfaction, you can achieve remarkable results. Despite its early stumbles, Walmart's tuition has achieved an outstanding learning curve and is delivering noteworthy bottom-line results today. In 2012, Walmart's international division, consisting of 6,242 stores (57 percent of Walmart's total), reported:

- $135.2 billion sales! That's 84 percent higher than Target's total sales. Don't forget, Target is considered to be Walmart's toughest domestic rival.

- $6.7 billion operating profit! That's over six times what Carrefour, the world's second-largest retailer, earned from operations.

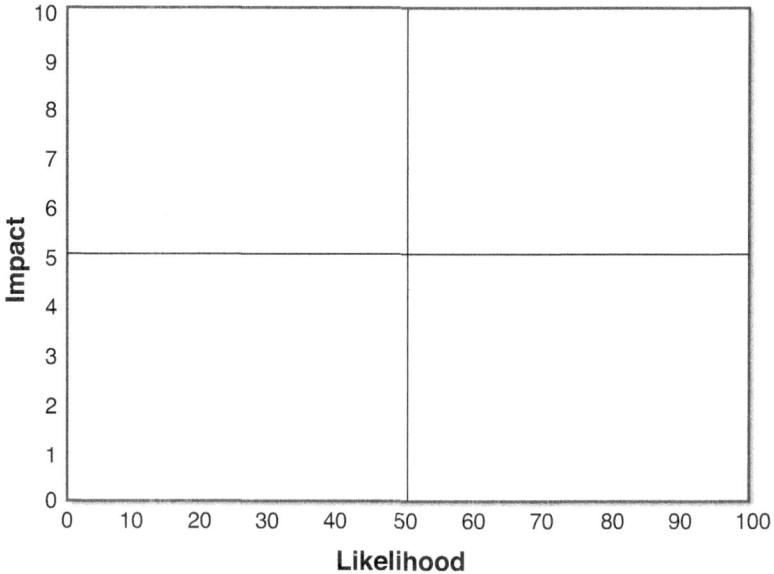

Figure 4-6 Risk assessment matrix

Conclusion

The difference between victory and defeat is often inglorious infrastructure. Almost nobody seems to give infrastructure its due—perhaps because it takes hard work, patience, and a willingness to work out of the spotlight to build a winning infrastructure. The sporting world provides an apt example of the opportunity and challenge found in infrastructure design and deployment. In 2002, the U.S. men's basketball team, made up of some of the world's finest athletes, placed a disappointing sixth in the FIBA World Championships. The U.S. team provided a disastrous encore in the 2004 Olympics as it lost three games on its way to a bronze-medal finish. The three losses were more than every preceding U.S. team had lost in all previous Olympic tournaments *combined*. USA Basketball had seen enough and was ready to blow up its approach to building a competitive team.

Jerry Colangelo was tasked with the overhaul. After carefully assessing the factors that had contributed to the team's subpar performance, Colangelo articulated a need for the new *infrastructure* he was about to put into place:

> Look, I was appalled by the status of USA Basketball. It needed a new *infrastructure*. In the past, teams were thrown together for a couple of weeks, and that was OK since the gap with the rest of the world was big enough it didn't matter. But

if you look at the successful international teams, they've been together for five or 10 years, so you get that continuity and teamwork. The first thing I wanted to establish was a real national team, not just an all-star team.[87]

With diagnosis in hand, Colangelo's first step was to find a coach who believed in the team-first concept. Mike Krzyzewski, the head coach at Duke University, was hired. Next, Colangelo and Krzyzewski defined job descriptions (i.e., roles and responsibilities) for each position. One-on-one interviews followed. The goal: to assess each player's commitment to the team and to the mission of restoring USA's position of dominance in international basketball. Athletic ability and talent were less important than attitude and fit.

With a blueprint for identifying the right players, Colangelo proceeded to change the rules for participation on the team. He insisted that each player commit to playing for Team USA for three years—a length of time that included playing in much-less prestigious tournaments. Krzyzewski explained, "You do not select a team, you select a group of people and then work together to develop into a team. In other words, teams don't instantaneously become, they evolve. To do so, you need time, goals, and competition."[88] The newly designed process, and the team it built, produced gold in the 2008 Olympics.

Infrastructure development can be a powerful weapon in your competitive arsenal. Amazon has shown how the right infrastructure can change customer habits—and expectations. As a result, Amazon is a retail darling and enjoys an envied image, stakeholder support, and persistent growth. However, for every Amazon, there are dozens of wannabes—companies that talk about, but do not invest in the infrastructure needed to deliver a truly remarkable customer experience. The wannabes have yet to figure out that success requires combining a variety of strategies and tools such as supply chain alliances, postponement strategies, information technologies, and bricks and mortar to deliver outstanding cost and availability performance.

Endnotes

1. "Which Brand Has the Best Advertising? Honda." 2004. *Creative Review* (September 2).

2. Dobele, A., Toleman, D., and Beverland, M. 2005. "Controlled Infection! Spreading the Brand Message through Viral Marketing." *Business Horizons* 48(2): 143–149.

3. "The New Honda Accord—Isn't It Nice When Things Just... Work?" 2003. *The Guardian*. Retrieved September 20, 2013, from http://www.theguardian.com/media/2003/may/09/2

4. Letts, Q. 2003. "Lights! Camera! Retake!" *The Daily Telegraph,* April 13. Retrieved September 20, 2013, from http://www.telegraph.co.uk/news/worldnews/europe/france/1427400/Lights-Camera-Retake.html

5. Boyer, T. 2003. Practical Motion. *Millimeter* (August 1).

6. Ibid, 4

7. Kite, J. 2009. "Spirit of the World Many Voices/One Truth." *Interspirit.net.* Retrieved September 23, 2013, from http://interspirit.net/spirit/spirit.cfm?ref=100572

8. Banjo, S., and Ziobro, P. 2013. "After Decades of Toil, Web Sales Remain Small for Many Retailers." *The Wall Street Journal,* August 28:A1.

9. MWPVL International Supply Chain. Amazon.Com Distribution Network. Retrieved June 15, 2013, from http://www.mwpvl.com/html/amazon_com.html

10. Gara, T. 2013. "Amazon Losing Its Price Edge." *The Wall Street Journal,* August 20:B2.

11. Ibid

12. Ibid, 8

13. Banjo, S. 2013. "Walmart's E-Stumble with Amazon." *The Wall Street Journal,* June 19:B1.

14. Ibid

15. Ibid

16. Stalk, G., Evans, P., and Schulman, L. 1992. "Competing on Capabilities: The New Rules of Corporate Strategy." *Harvard Business Review* 70(2):57–69.

17. Lepore, M. 2011. "Here's How Walmart Became the #1 Grocery Store in the Country." February 11. Retrieved September 20, 2013, from http://www.businessinsider.com/walmart-biggest-supermarket-2011-2?op=1

18. Ibid, 13

19. Walmart Stores, Inc. 2013. Where in the World Is Walmart? Retrieved September 20, 2013, from http://corporate.walmart.com/our-story/locations

20. Walmart Stores, Inc. 2013. Unit Counts & Square Footage. Retrieved September 20, 2013, from http://stock.walmart.com/financial-reporting/unit-counts-square-footage

21. MWPVL International Supply Chain. 2013. The Walmart Distribution Center Network in the United States. Retrieved September 2, 2013, from http://www.mwpvl.com/html/walmart.html

22. Ibid, 21

23. Walmart Stores, Inc. 2013. Walmart U.S. Logistics. Retrieved September 20, 2013, from http://corporate.walmart.com/our-story/our-business/logistics

24. Ibid, 21

25. Ibid, 23

26. Ibid, 21

27. Ibid, 23

28. Walmart Stores, Inc. 2013. Walmart Facts. Retrieved September 20, 2013, from http://news.walmart.com/walmart-facts

29. Walmart Stores, Inc. 2013. Apply to Be a Supplier. Retrieved September 20, 2013, from http://corporate.walmart.com/suppliers/apply-to-be-a-supplier

30. Souza, K. 2012. "Wal-Mart Suppliers Create Momentum for NWA Economy." *The City Wire*. Retrieved September 20, 2013, from http://www.thecitywire.com/node/24530#.Uj0nOoKAGF1

31. Walmart Stores, Inc. 2013. *Walmart Annual Report*. Bentonville, AR.

32. Fawcett, S. E., Ellram, L., and Ogden, J. 2007. *Supply Chain Management: From Vision to Implementation*. Upper Saddle River, NJ: Prentice Hall.

33. Ibid, 13

34. Ibid, 13

35. Ibid, 13

36. Spiekman, P. 2000. "New Victories in the Supply-Chain Revolution." *Fortune*: T208C-T208HH.

37. Maloney, D. 2008. "Whirlpool's Delicate Cycle." *DC Velocity*. Retrieved September 2, 2013, from http://www.dcvelocity.com/articles/20080901verticalfocus/

38. Kumar, N. and Steenkamp, JB. 2007. *Private Label Strategy: How to Meet the Store Brand Challenge*. Harvard Business Publishing, Boston.

39. McArdle, M. 2012. "Why Can't Walmart Be More Like Costco?" *The Daily Beast*, November 26. Retrieved September 2, 2013, from http://www.thedailybeast.com/articles/2012/11/26/why-can-t-walmart-be-more-like-costco.html

40. Rohwedder, C., and Kesmodel, D. 2009. "Aldi Looks to U.S. for Growth." *The Wall Street Journal,* January 13. Retrieved September 2, 2013, from http://www.online.wsj.com/article/SB123180518793975423.html

41. Bustillo, M. 2010. "Wal-Mart See Small Stores in Big Cities." *The Wall Street Journal,* October 13. Retrieved September 2, 2013, from http://www.online.wsj.com/article/SB10001424052748703673604575550243762557882.html

42. Ibid, 32

43. Tate, W. 2010. "A Primer on Sourcing and Procurement in an Integrated Supply Chain." *Supply Chain Management Review,* October 21. Retrieved September 23, 2013, from http://scmr.com/article/a_primer_on_sourcing_and_procurement_in_an_integrated_supply_chain; Ibid 32

44. Kraljic, P. 1983. "Purchasing Must Become Supply Management." *Harvard Business Review* 61(5):109–117; Gottfredson, M., Puryear, R., and Phillips, S. 2005. "Strategic Sourcing: From Periphery to the Core." *Harvard Business Review* 83(2):132–139.

45. McGrath, M., and Hoole, R. 1992. "Manufacturing's New Economies of Scale." *Harvard Business Review* 70(3):94–102.

46. Laseter, T. M. 1998. *Balanced Sourcing.* San Francisco: Jossey-Bass Publishers.

47. Schwartz, E. 2004. "Wal-Mart Promises RFID Will Benefit Suppliers." *Infoworld.* Retrieved September 4, 2013, from http://www.infoworld.com/t/data-management/wal-mart-promises-rfid-will-benefit-suppliers-160

48. Walmart Stores, Inc. 2009. *Wal-Mart 2009 Global Sustainability Report*: Walmart Stores, Inc.

49. Horick, R. 2008. "Category Captains: Who's in Charge?" *Vanderbilt University Owen Graduate School of Management.* Retrieved September 23, 2013, from http://www.owen.vanderbilt.edu/faculty-and-research/vanderbilt-business-inbrief/category-captains-whos-in-charge.cfm

50. Ibid, 30

51. "Walmart: Loyalty Cards? We Don't Need No Stinkin' Loyalty Cards!" 2013. *Grocery & Retail News,* March 19. Retrieved September 4, 2013, from http://couponsinthenews.com/2013/03/19/walmart-loyalty-cards-we-dont-need-no-stinkin-loyalty-cards/

52. "Self-Scan's Biggest Benefit: Convenience, Cost Savings, or Coupons?" 2012. *Grocery & Retail News,* September 4. Retrieved September 4, 2013, from http://couponsinthenews.com/2012/09/04/self-scans-biggest-benefit-convenience-cost-savings-or-coupons/

53. Fawcett, S., and Magnan, G. 2001. *Achieving World-Class Supply Chain Alignment: Benefits, Barriers, and Bridges.* Phoenix, AZ: National Association of Purchasing Management; Ibid 32

54. Ibid, 36

55. Ibid, 13

56. Ibid, 13

57. Ibid, 13

58. Senge, P. 2006. *The Fifth Discipline: The Art and Practice of the Learning Organization.* New York: Doubleday.

59. Ibid

60. Ibid

61. Ibid, 31

62. Ibid, 31

63. Ibid, 16

64. Ibid, 8

65. Ibid, 8

66. Ibid, 13

67. Banjo, S. 2013. "Wal-Mart: A Pro in Physical-Store Retail Logistics." *The Wall Street Journal,* June 19:B2.

68. Ibid, 31

69. Ibid, 13, 67

70. Ibid, 58; Churchman, C. 1968. *The Systems Approach.* New York: Dell Publishing.

71. Ibid, 13

72. Peters, T. 1986. "What Gets Measured Gets Done." *tompeters Column Archives,* April 28. Retrieved September 23, 2013, from http://www.tompeters.com/column/1986/005143.php

73. Thomas, K. 1992. "Conflict and Negotiation Processes in Organizations." In D. Dunnette, and L. M. Hough (Ed.). *Handbook of Industrial and Organizational Psychology* (Vol. 3). Palo Alto, CA: Consulting Psychologists Press.

74. Henkoff, R. 1994. "Delivering the Goods." *Fortune* (November 28):64–78.

75. Ibid

76. Walmart Stores, Inc. 1993. *Walmart Annual Report.* Bentonville, AR.
77. Friedland, J., and Lee, L. 1997. "The Wal-Mart Way Sometimes Gets Lost in Translation." *The Wall Street Journal,* October 8:A1, A12.
78. Ibid
79. Ibid
80. Michaels, D., and Sanders, P. 2009. "Dreamliner Production Gets Closer Monitoring." *The Wall Street Journal,* October 7.
81. Lunsford, J. 2007. "Boeing Scrambles to Repair Problems with New Plane." *The Wall Street Journal,* December 7:A1.
82. Gate, D. 2011. "Boeing Celebrates 787 Delivery as Program's Costs Top $32 Billion." *The Seattle Times,* September 24. Retrieved September 23, 2013, from http://seattletimes.com/html/businesstechnology/2016310102_boeing25.html
83. "Supply Chain Risk Management: Building a Resilient Global Supply Chain." 2008. *Aberdeen Group,* July 31. Retrieved September 23, 2013, from http://aberdeen.com/Aberdeen-Library/4185/RA-global-supply-risk.aspx
84. 53% of Supply Chain Leaders Say Disruptions Now More Costly. 2013. *@Risk,* June 21. Retrieved September 7, 2013, from http://atrisk.net/53-of-supply-chain-leaders-say-disruptions-now-more-costly/; the Rising Tide of Supply Chain Risks: How Risk Managers' Role and Responsibilities Are Changing." 2008. *Marsh,* April 15. Retrieved September 23, 2013, from http://usa.marsh.com/NewsInsights/ThoughtLeadership/Articles/ID/539/Stemming-the-Rising-Tide-of-Supply-Chain-Risks-How-Risk-Managers-Roles-and-Responsibilities-Are-Changing.aspx
85. Ibid, 84 @Risk
86. Ibid, 84
87. Rhoads, C. 2008. Questions for Jerry Colangelo. August 22. Retrieved May 5, 2012, from http://online.wsj.com/article/SB121939315839463251.html
88. Krzyzewski, M. 2009. *The Gold Standard Building a World-Class Team.* New York, NY: Hachette Book Group.

5

IMPLEMENTING AN ENABLING TECHNOLOGY STRATEGY

Opening Story: The Allure of Technology

October 25

At the Espresso Machine

"Good morning, Diane," Paul said smiling. "It certainly is a beautiful day."

Diane's glance was chilling as she replied, "It's raining and it's nasty cold. You call that beautiful?"

"Not exactly," Paul admitted, "but I'll be leading our discussion on IT in the task-force meeting later this morning. What more could I ask for—except a hot cup of cocoa? By the way, Diane, do you mind if I ask a touchy question?"

"I suppose I'm game. What's on your mind?" Diane queried.

"That was a gutsy move to pull the five-year SC technology plan. What were you thinking that led you to opt out of your meeting with the strategic planning committee 24 hours before your presentation?"

"You're right, Paul. That was a tough decision. We had put in a ton of time and effort building the plan. But, the moment I asked David to reimagine our order fulfillment capabilities, our five-year technology plan became obsolete. Better to acknowledge that right up front. Now that we know we need to rethink fulfillment, we need to ask if there are there any 'new' technologies that could enable us to keep Monster, Inc., and other customers happy. Besides, as we redefine the fulfillment process, we will know far better what technologies we need. Although new technologies may enable a new and better process, the process will define the technology. We can't afford to put the cart before the horse. I'm counting on you to help us get this figured out—starting with your presentation later this morning."

Two Hours Later—In the Conference Room

"Good morning everyone," David said. "I'd like to formally welcome Paul, our VP of IT, to the task force. These days, IT affects almost everything we do in fulfillment. We all agree that IT is great—when it works! But, it doesn't always work the way we think it should or thought it would. Having Paul on the team should help us get the redesign process right. Paul, you're up. What do we need to know about IT to build a world-class fulfillment capability?"

"Good morning. I'm delighted to join the team. Working together will help us better grasp what's possible. It will also help us avoid some costly pitfalls. With that in mind, I'd like to take out some trash. By that, I mean that I'd like to dispel some misperceptions about IT and its role in helping DWC really meet our customers' needs at a whole new level. I have four thoughts I'd like to share:

1. We speak different languages. It has been said that IT is from Venus; the rest of the company is from Mars. I like to say it this way, 'You can't cut cheese so thinly that it only has one side.' The simple truth is that we may work for the same company, but we live in different worlds and face very different challenges—or at the very least, different priorities. If we learn to understand each other, we can avoid the miscommunications and mistrust that emerge when we talk past each other instead of with each other. That's one reason I'm grateful to be on the team. The time invested now will save us both heartache and money down the road.

    Please don't be disappointed if you tell me, 'We want it this way' and I respond, 'We can't do it that way.' My team will do its best to share with you the costs, benefits, and risks of doing something your way. Then we'll try to give you a couple of other options that will meet your performance requirements better. Of course, we will document their costs, benefits, and risks. Then we'll have to make the decision. Likewise, I've heard the scuttlebutt in the hallway. You see our methodologies as hoop-jumping bureaucracy. But, without the strict methods, when we throw the switch on a new system, I can guarantee you that the system won't work.[1]

2. IT is no cure all. We've all heard about the classic IT failures. Hershey's ERP implementation was a disaster, keeping Hershey from delivering $100 million worth of Kisses and Jolly Ranchers for Halloween in 1999.[2] Walmart's push to have its top 100 suppliers using RFID tags was really big news—both when it was

announced and when it was called off.³ Technology implementations are tough for several reasons. Sometimes, like in the case of RFID, the first adopters get out ahead of the technology. That is a risk of being on the cutting edge. Always, new technologies require that companies, and the people who run them, change the way they do business. Let me assure you, there are no technology silver bullets! But, there are plenty of people willing to sell you one. If we buy a panacea, we are buying trouble. There is no way to get around the hard work of process change.

3. IT does have a real value-added role. The research is clear, IT-savvy companies are 21 percent more profitable than their nonsavvy counterparts.⁴ IT not only offers opportunities to reimagine processes, but it can also help standardize and digitize core processes, offering tremendous efficiency advantages.⁵ At the same time, when the technology architecture is integrated, companies bring products to market faster—they are simply more agile. A well-designed digital platform also enables companies to collaborate more effectively with partners up and down the supply chain. As you can see, IT-savviness is not about the IT department; it is about all of us thinking digitally—constantly asking, 'Can IT help us standardize or innovate this process?'

4. IT is an enabler, but people make the difference. One of the great IT ironies is that we invest in the wrong things.⁶ IT is an enabler; it is not a solution. IT can provide the data, but our people need to provide the insight. IT can help us see things in new ways, but we must do the ideating and innovating. If we get the balance wrong, if we try to substitute IT investment for good thinking and good management, no amount of 'Big Data' will help us win tomorrow's competitive battles.

If we forget any of these four points, we might be the latest Fortune 500 company to grace the front page of the *The Wall Street Journal* for botching a big IT implementation. I'm sure we can avoid that fate. More important, I'm confident IT can help us establish a customer-winning fulfillment capability. I hope I've given us something valuable to think about as we embark on the reimagination process."

Consider as you read:

1. Based on your experience, which of the four points Paul makes is the most common pitfall to managing IT for logistical process reimagination? Is there a common thread among these points?
2. Why was it gutsy for Diane to postpone the five-year IT plan presentation to the steering committee? Was it the right thing to do? Why?
3. What approaches/tools can be used to mitigate the downside risk of a failed IT implementation?

Implementing an Enabling Technology Strategy

"Where is the wisdom we have lost in knowledge? Where is the knowledge we have lost in information?"

—T. S. Eliot

The information revolution gave life to modern supply chain management. Even so, most companies still struggle to harness the power of technology. Why, you ask? The answer lies in the allure of technology—an allure you probably know firsthand. Have you ever been seduced—by technology? If so, you share a common experience with most corporate decision makers. You know the deal. A new high-tech gadget—for example, the iPhone or iPad—comes to market and you've got to have one. Your quest is driven by dual passions. First, you *fear* being left behind. If everyone else is buying a smartphone, you've got to have one, too—even if you don't know how to use it or you won't fully use its capabilities. Second, you *imagine* that buying the latest technology will make your life better. Although it might, experience has taught you that technology doesn't solve all of your problems. Misused, technology might even cause you unimagined grief.

Ponder, for a minute, these two motivations: fear and fantasy. They are powerful and are fully embedded in your technology choices. Thus, technology—unlike other resources—often causes rational decision makers to lose sight of strategy, dismiss budget discipline, and misplace common sense. How does this happen? The choice to invest in technology elicits both the fear of loss and a hope for gain. No one wants to fight tomorrow's competitive battles with yesterday's technology.[7] No one ever got fired for winning with technology. Even as you pursue technology as part of a defensive arms race, you envision its possibilities. With your imagination piqued, you begin to see fantastic new technology-enabled capabilities—opportunities to do things better and faster than ever before. Then reality reappears. Your technology-inspired visions outrun your ability to reengineer processes and redefine supply chain relationships. Consider, for example, the following story about how technology was destined to change the order fulfillment process. A brief quiz will follow.

> *It is nine o'clock on a Saturday morning in May [date deleted]. The time has arrived for Steve Robinson to part with $50,000 for a new car. For months, he has been thinking about retiring his six-year-old Chevy Blazer, decidedly obsolete by the standards of the day. The last straw that spurred him to action was when one of his teenage children said several days ago that she was ashamed to be seen riding in such an old vehicle.*

> *After heated discussions, the Robinsons have agreed on a sporty new Ford Explorer as their vehicle of choice. Over recent years, the Explorer has evolved from its original utilitarian design and some of the latest models have all the features of a top-of-the-line luxury vehicle.*

Before leaving home to order the Explorer, Steve dials to channel 469 on his Web television, which he still thinks of from time to time as just a television with cable. He notifies his local bank that he plans to purchase a car within the next seventy-two hours, reserves $50,000 for payment when he makes the purchase electronically, and receives approval for the funds transfer. The entire transaction takes only 14 keystrokes. Steve still uses the keyboard attached to the television monitor—a fact that does not escape his son, who constantly admonishes him to be "up-to-date" and use either the remote or the voice recognition and response module to work with the monitor.

An hour later, Steve arrives at Sam's Club to make the final decision. He pauses briefly at a large overhead sign that proclaims "Welcome to Sam's Virtual Reality Car Showroom" and then proceeds to the sign-in station. After he has entered his universal identification number and bank credit line password, the showroom's system instantly identifies him and grants him access, recording the bank approval for $50,000.

The showroom has four sets of seats where shoppers can experience, through the wonders of virtual reality technology, the look and feel of different car makes and models in a variety of driving environments. As he settles into one of the seats, Steve specifies to the system that he wants to try out a black, four-door 4x4 Explorer with all the options, including an in-car fax hooked into the onboard trip computer. Next, the system prompts him to request a test drive to challenge the sport utility vehicle and his ability to handle it under adverse weather and road conditions. Steve decides to take a virtual reality drive on an October day in New Hampshire, first under sunny conditions, then in the rain. Both the smooth, pleasant drive and the beautiful fall colors exceed his expectations. He is sold.

At the upper right-hand corner of the keyboard is a large red button labeled BUY. Pushing the button, Steve says good-bye to his $50,000 and triggers the process that will cause his new car to roll off the assembly line five days from today and be delivered to his house two days after that. Seven days seems an eternity, but he is glad to be making the purchase.

Leaving the showroom, Steve muses about the big red button. Why not voice recognition technology? he wonders, recalling his son's comment that everyone has been using it for quite a while. Concluding that the button is a gimmick to heighten the thrill of making the purchase, Steve turns his thoughts back to the Explorer and heads for home.

Virtual reality car showrooms? In a Sam's Club? Voice recognition in an interactive purchasing process? Secured electronic attachment of funds through a cable channel? A universal personal identification number complete with on-line personal information accessible anywhere with the use of a password? All this sounds futuristic...Or does it?

The reality is that the technology to do all of these things is here today.[8]

Time for the quiz: When was this story written? You may be surprised that Fred Kuglin shared this scenario in his 1998 book, *Customer-Centered Supply Chain Management*. Over 15 years later—despite the fax's demise and Web TV's rise—the story still sounds futuristic. If the technology was available in 1998, why aren't we using virtual reality to enhance our daily lives? Why aren't automakers building custom cars for seven-day delivery? Why hasn't technology led to widespread home fulfillment?

W. Brian Arthur, Stanford professor and expert on complexity theory and technology, provides the following explanation:

> One thing I've learned from looking at past technology revolutions is that the deeper the transformation, the more slowly it takes place. Electric motors became available around 1880. At that time all machines in a factory were powered by a single, lumbering steam driven motor. The new electric motors could each power a single machine—a weaving loom, for example—allowing flexibility and potential cost savings. But their proper use required redesigning factories and production processes, and that took more than 40 years. *Economic transformation is slow not because it requires new equipment but because it requires new—and often not obvious—ways to* **organize** *business.*[9]

What should you take away from this discussion? A few leaders, like Jeff Bezos at Amazon, grasp the nuance of technology-driven process reimagination. They recognize technology's potential, but they know change is hard. They therefore take a process-first approach, painstakingly setting expectations and taking the slow-but-steady steps to organize for success. By contrast, many managers, like the founders of Webvan (the $1-billion defunct home-delivery service that has been called the greatest flop in dot.com history[10]), take a technology-first approach. Seduced by what technology can do, they misjudge the pace of transformation, failing to reorganize the way business is done. They overlook employee and customer reluctance to change the way they work, shop, and live. They outrun their process capabilities. Ultimately, you need to remember that a valued, technology-enabled capability goes far beyond the technology's capability. It is always organizational in nature.

The Nature of Information-Technology Enablement

Companies enter technology's danger zone when managers focus more on what IT can do than on why IT is being adopted. As this happens, managers begin to see IT as *the* solution rather than as an enabler. The goal shifts from enabling a specific, value-added capability to implementing cutting-edge technology. This technology "creep" produces two negative effects:

- Managers are often blinded to the organizational issues that underlie successful IT implementations. As process change, training, and measurement are overlooked, managers put the cart (technology) before the horse (process design).

- IT implementations take longer than planned, exceed budget, fail to deliver the promised improvements, and are easily copied by rivals. Competitive advantage remains elusive.

To avoid technology creep, you need to define clearly what a valued information capability looks like—that is, how IT really enables better order fulfillment and customer service. From this perspective, your goal is simple: You want to use IT to quickly and inexpensively *share* accurate and relevant information with decision makers across your organization as well as up and down the supply chain. Effective *information sharing* can deliver remarkable operating improvements (see Figure 5-1). For example, IT-enabled *information sharing* allows managers to (1) substitute information for inventory, (2) take time out of value-added processes such as order fulfillment, (3) pursue process reengineering, and (4) collaborate more effectively with supply chain partners.

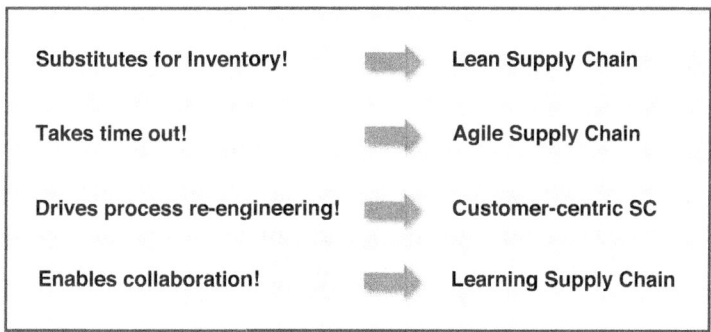

Figure 5-1 Benefits of an IT-enabled information-sharing capability

With so much value-creation potential on the line, a closer look at the nature of an information-sharing capability is merited. As Figure 5-2 depicts, an outstanding information-sharing capability consists of two distinct components: connectivity and willingness.[11] At most companies, the need for better connectivity—and the IT investments needed to achieve it—is clearly and widely perceived. By contrast, the call to invest in a culture of willingness to share critical, often sensitive information remains largely unheeded. Only slowly are managers beginning to recognize that the technological ability to connect does not equate to a willingness to share—nor does it lead to great information sharing. Connectivity and willingness are equally needed.

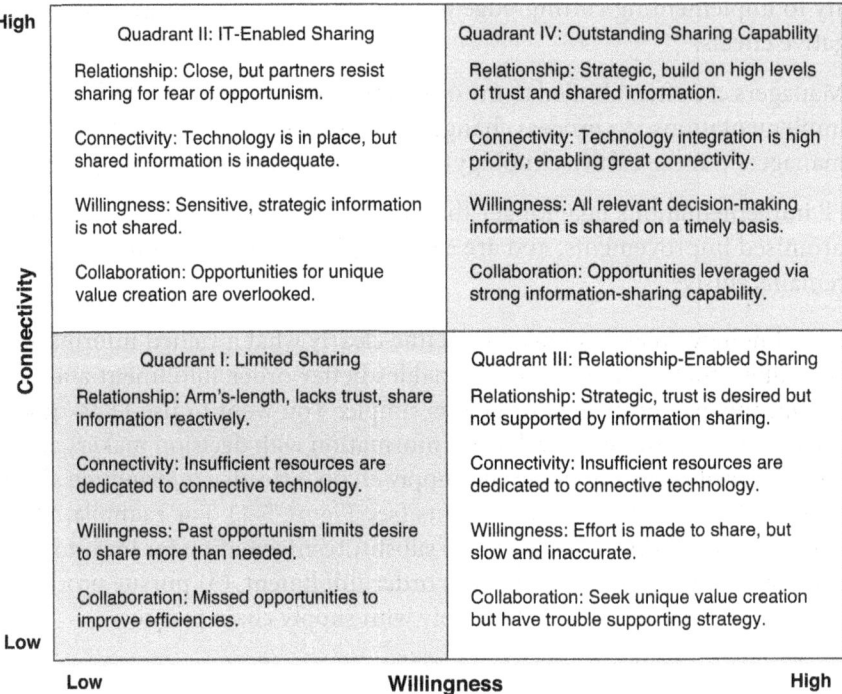

Figure 5-2 Components of an information-sharing capability[12]

A Closer Look at Connectivity

Connectivity is defined by your company's ability to use IT to collect, analyze, and disseminate needed decision-making information. At most firms, connectivity is widely viewed as critical to effective supply chain management and is directly correlated with well-managed technology investments. Over the past decade, companies have invested billions of dollars in IT. The investments have paid off in much higher levels of connectivity. Companies are learning how to use the Internet, enterprise systems, and Big Data tools (i.e., data capturing, storage, and analytics) to build better, more reliable processes; effectively analyze decision-making tradeoffs; and work more effectively across organizational boundaries.

As the following examples illustrate, IT innovations and enabling investments are taking place at almost every step from concept to delivery in the order fulfillment process:

- **Design**—Common manufacturing databases are being used to standardize parts lists and seek aggregation opportunities. Combined with integrated computer-aided design (CAD) systems, buyers and suppliers can actively collaborate on new product development—regardless of geographic location. The 24-hour design day is now a reality for many global companies. Better collaboration and global teaming are dramatically reducing concept-to-market cycle times and development costs.[13]

- **Supply management**—Web catalogs enable process standardization, reducing the time and effort expended by buyers as well as end users on standard buys. They also enhance internal compliance, reduce relationship uncertainty, and lower transaction costs. Perhaps most important, once the catalog is set up, more time is available to work on strategic initiatives with a higher potential ROI.[14]

- **Production control**—Sensors embedded inside machines track operations. If the wrong material is used or the power goes down, the sensors alert a remote operator via text. Using her iPad, the operator can pull up schematics and real-time data, intervening from afar—if necessary.[15] Smart machines monitor the minutest production details, including deviations in how many times a screw has been turned. If the screw is turned 12 times instead of the required 13, production is halted. Commenting on how these machines are redefining quality control, Ed Magee, general manager at a Harley production plant, marvels, "You can solve a problem before it results in machine downtime."[16]

- **Forecasting and inventory control**—Leading companies use Web portals to give partners access to up-to-date order histories, sales trends, and rolling forecasts. Walmart's Retail Link enables suppliers to monitor sales on a real-time, store-by-store basis. The supplier is expected to use this data to manage replenishment, forecast future demand, and develop merchandising plans. By sharing data, Walmart enlists suppliers to ensure the right product is on the shelf for customers to buy. Of note, the suppliers pay far better attention to their own products' availability and flow-through rates than Walmart could.[17]

- **Warehousing and material control**—RFID is used to track pallets and cases through the fulfillment system all the way into the back door of a retail store at a higher level of precision than is possible using bar codes. With the launch of the Fusion razor, P&G used RFID to track case shipments to 500 stores, achieving a 92 percent on-shelf availability in only 3 days—a level of store penetration normally reached only after 14 days.[18] Data from RFID-tagged display cases later led P&G to pull the plug on the display-tagging experiment when P&G determined that poor in-store execution meant the displays were too often sitting unused in the stores' back rooms.[19] Although the idea of item-level tagging has dissipated in the CPG-to-retail market, it is alive and well in higher-margin fashion retailing.[20]

- **Transportation**—Onboard computers and global positioning systems (GPS) have transformed transportation into a technology industry. Onboard computers help analyze and improve driver performance and safety even as they support sustainability initiatives. GPS makes it possible to track real-time status and location, enabling better routing and faster, more flexible expediting. The next step: autonomous trucks? Although it may sound surreal to envision tractor-trailer rigs running down the interstate without drivers, Caterpillar is already testing satellite-guided, self-driving mining trucks in an iron-ore mine in Australia.[21] (Google and Mercedes have already tested self-driving cars.) Trucking companies may be on the verge of alleviating one of their biggest headaches—a persistent driver shortage.

No doubt, the order fulfillment process of the future will look and act very different from what you manage today. However, many challenges remain to achieving the fully connected supply chain. In fact, many companies continue to find it frustratingly difficult to integrate information applications—both within the firm and across the supply chain. The cost—in dollars, time, and understanding—continues to impede ERP and other state-of-the-art IT implementations. Further, trading partners often pursue different priorities and are at different stages in their connectivity quest. As a result, they may not be able to connect as equal partners. This is especially true for smaller, capital-constrained customers and suppliers. By its very nature, connectivity cannot be achieved without the active participation of supply chain partners.

A Closer Look at Willingness

Willingness describes your company's cultural predisposition to share the information colleagues need to make great (timely and correct) decisions. Willingness is a behavioral trait—one that is often overlooked in the quest to build an information-sharing capability. Although connectivity makes it possible to share information quickly and efficiently, your culture of willingness (or lack thereof) determines what and how much useful information is shared. For example, most firms are now willing to share up-to-date historical sales, rolling forecasts, and scorecard information. Not only are they employing technology to do so, but they are also engaging in more interorganizational teams and senior-manager interaction to enhance sharing.

What they are not sharing is sensitive or so-called "proprietary" information—at least not at the levels partners desire. Why do partners want access to proprietary information? So they can make necessary investments to support the relationship and enhance value creation over the long haul. Yet, technology road maps, market-entry plans, and new product strategies are often off the table or shared on a selective need-to-know basis. (As you might imagine, the two sides of many relationships disagree on what "need to know" really means.) The result: A lack of willingness to share "all relevant" information is viewed by many managers as a greater hindrance to value creation and customer satisfaction than the more frequently publicized connectivity challenges.

Given companies have invested a ton of money in IT to make sharing possible and almost everyone agrees that information sharing is the lifeblood of efficient and agile supply chains, why is a behavioral trait like willingness holding up progress? The answer is quite simple. Ask yourself, "What is information?" If you answered, "Information is power," you identified the number-one response to the question. You also solved the mystery. Sharing information openly empowers decision makers across the supply chain to make better decisions and create more value—both good outcomes for the entire supply chain. However, open sharing places those who share sensitive information in a vulnerable position.[22] Someone in the chain might inappropriately use the information for individual gain. Fearing opportunism, many managers hoard rather than share information. The potential for global gain is not great enough to overcome the fear of personal loss, keeping companies from realizing the full benefits expected from IT investments.

To summarize, Figure 5-3 puts performance numbers to the quadrants in the connectivity-willingness matrix to highlight that companies forfeit both efficiency and customer satisfaction if they fail to build a robust information-sharing capability. Data regarding efficiency and customer satisfaction were collected from almost 1,400 companies to see if combining connectivity and willingness really leads to better performance. Each company was assigned to a quadrant in the 2×2 matrix and mean performance scores were compared (1=low; 5=high). Companies in Quadrant IV: High Connectivity-High Willingness performed at significantly higher levels than companies in the other three quadrants. Not surprisingly, companies that neglect both connectivity and willingness lag dramatically behind. The bottom line: You maximize your return on investment when you invest in IT to create connectivity and cultivate a culture that enables people to be willing to share.

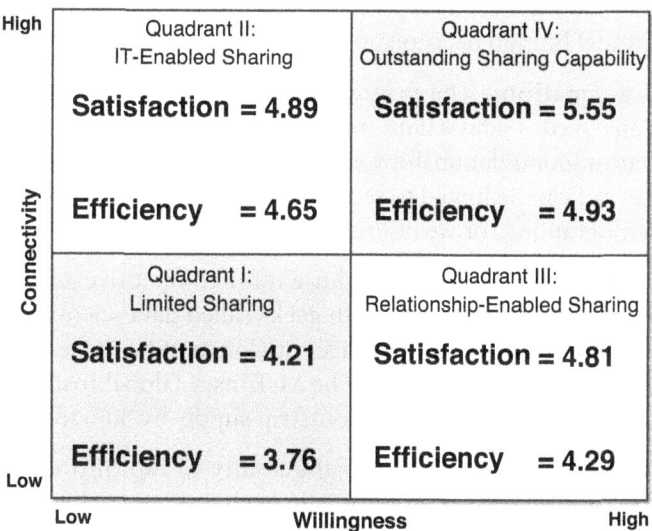

Figure 5-3   The performance impact of an information-sharing capability[23]

Moving Toward Information Enablement

Despite the proven opportunity to use IT to reengineer processes and enable unique capabilities, few companies have fully leveraged IT's potential. A McKinsey study of early reengineering efforts revealed that 14 out of 20 (70 percent) had failed to achieve targeted goals.[24] A key finding of the study was that companies tended to focus on isolated initiatives (i.e., local process improvements), forgetting about broader systems effects. A decade later, Michael Hammer, the father of reengineering, lamented that fewer than 10 percent of large corporations had used IT to truly reimagine processes.[25] Companies had failed to change the way work is done. The question naturally arises, "Why is IT enablement so difficult?" You need to be aware of at least two issues that plague IT implementations: (1) Investment patterns minimize IT's power to make a difference, and (2) many companies fail to follow a proven path to success.

Understanding Investment Patterns

How a company invests in and manages IT implementations has a strong influence on whether or not the investment leads to a distinctive capability. Figure 5-4 depicts a basic IT-enablement hierarchy. Four capability levels exist:

- **Level 1: Data**—Data is the raw material used to make decisions and to build an information-sharing capability. Big Data connotes today's challenge. Smart machines, bar codes, RFID, electronic sensors, click-through counts on the Internet, social media content, and mobile devices provide gargantuan amounts of information about customers, products, and processes—all of which can be harvested to model human decision-making and operating system behavior.[26]

- **Level 2: Information**—The primary data collected at Level 1 becomes useful when it is analyzed. Today's data analytic tools provide immense power to model data so that unique relationships are more visible.[27] Further, primary data collected at Level 1 can be linked to secondary data sources such as census numbers, Nielsen demographics, or weather statistics, to derive unique insights.[28]

  To really make sense of the data and use it for competitive advantage, managers are learning how to work closely with geeks called data scientists. Unfortunately, as Jesse Harriott, chief analytics officer at Constant Contact, Inc., notes, there's "an acute skill shortage out there." The McKinsey Global Institute estimates that the demand for data scientists will "outstrip supply by 50% to 60% by 2018."[29]

- **Level 3: Knowledge**—Knowledge—the ability to tap into the experience and insight of the entire organization (or supply chain)—is enabled when technology makes local skills, talent, and understanding widely available. 3M employed an early knowledge management system to help employees find the right talent to use on product development teams. In an early example, an avid fly fisherman—frustrated by the sinking properties of existing fly lines—sought out polymer experts within 3M's system. The right idea leveraged with the right know-how led 3M into the fly-fishing market.

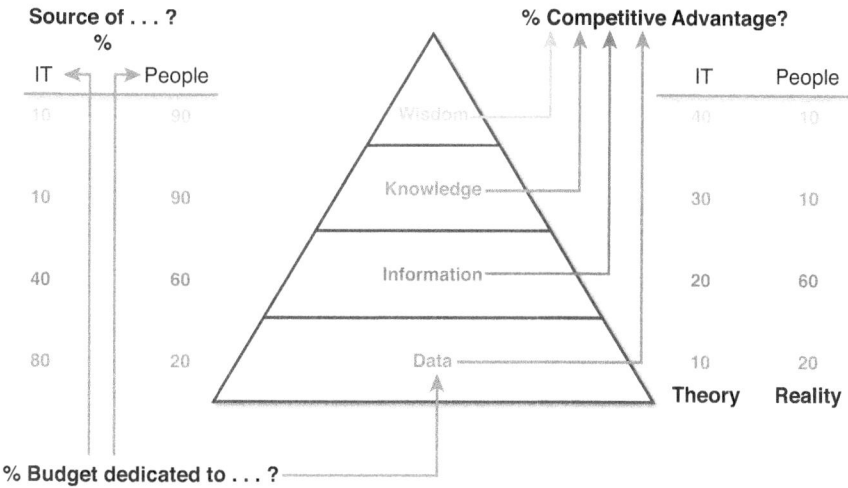

Figure 5-4 Investment patterns related to IT enablement[30]

More recently, 3M has taken the idea one step further, using the Internet to conduct an "InnovationLive" experiment designed to capture ideas for "Markets of the Future." Over 1,200 3Mers from 40 countries participated, generating 736 ideas and ultimately identifying nine untapped new markets.[31] P&G has taken crowdsourcing/open innovation to its entire supply chain to invite wide open ideation. The result: By 2010, P&G got over 42 percent of new product ideas from outside the organization (up from only 10 percent at the turn of the millennium).[32]

- **Level 4: Wisdom**—Wisdom is the highest—and rarest—form of IT-enabled capability. Wisdom is a technology-enabled organizational ability to learn. One measure of wisdom is how long it takes for an idea (e.g., process improvement or new product) to be disseminated across an entire system (e.g., firm or supply chain). For years, perhaps Walmart's most overlooked strength was its ability to drive learning across its network. Consider the following examples:

  - A Walmart associate in Arizona observes that Hispanic customers are asking for a specific kind of cookware known as a caldero. She shares this insight during a morning meeting and calderos are soon on the shelf.[33]

- A display at Sam's Club fails to stand up straight, allowing a piece of furniture to slide precariously on the rack, potentially endangering customers. One of the workers identifies the problem and within the workday, the display is fixed at all Sam's Club stores across the country.

- A customer shared the following experience:

  I telephoned the Bentonville, Arkansas, headquarters of Walmart to complain about its store in La Plata, Argentina. The switchboard immediately rang the vice president of international operations, who picked up his own phone. He thanked me for calling, asked detailed questions about my dissatisfaction, and inquired whether I was willing to repeat my story for his Latin American VP. He transferred me straight away, and an even more detailed conversation followed. Then I was asked if I would be willing to talk with the Argentinian store manager if he called me. Ten minutes later my phone in Connecticut rang.

  On my next trip to Argentina, a year later, the store had been transformed. No wonder Walmart is the world's largest retailer.[34]

These stories reveal the "secret" formula that underlies a wisdom capability:

Observe + Ask + Analyze + Act = Organizational Learning

Observe, ask, and act are inherently human skills, unleashed or held captive by an organization's culture.[35] Analyze, however, is where today's Big Data technology has the potential to rewrite the rules of the game. Indeed, with the ability to dig deeper and see "more uniquely," people might start asking questions they thought could never be answered. Storytelling—a quintessential wisdom capability that creates the motivation and disseminates the blueprint for innovation—may be forever changed.

Now, with this background, consider the questions raised by Figure 5-4. Remember, your goal is direct investment to develop capabilities that deliver hard-to-copy customer value. First, "What is the source of each of these capabilities?" Each capability depends on people and technology working well together. Although a data capability is primarily technology intensive, people define what is collected, stored, and disseminated. The design process is management's responsibility. You might, thus, describe a data capability as 80 percent technology and 20 percent people. An information capability relies more on analysis and interpretation to make good, timely decisions. You would likely attribute much more importance to people—perhaps as much as 60 percent. The sources of knowledge and wisdom are even more people dependent. Technology is the conduit that enables the sharing of talent and ideas, representing perhaps as little as 10 percent of their source.

Second, "What percent of competitive advantage arises from each capability?" Remember in Chapter 2, "Fulfilling Orders: The Nature of Modern Order Cycle Management," we discussed value creation in terms of cost, quality, delivery, innovation, and responsiveness. You achieve competitive advantage when you outperform your rivals in these areas. In *theory,* advantage grows as you move from data to wisdom. Data provides limited advantage—perhaps only 10 percent. Data by itself does not change what your company does or how it competes. Further, a data capability can be easily copied—if your firm has the money to buy the technology and installation advisement. Information provides customer insight and is used to develop efficient processes and attractive products. However, an information capability should be accessible to most companies. Thus, information accounts for a modest 20 percent of advantage. Knowledge helps you leverage talent and relationships. As an organizational capability, knowledge routines are more difficult to copy and should yield greater advantage (say 30 percent). Wisdom, or learning, drives continuous improvement and differentiation. As a higher-order, culturally embedded routine, wisdom should be very hard to copy or cultivate, providing you the greatest (say 40 percent) advantage. Regrettably, theory and experience diverge. The reality is that information is the core driver of most firm's IT-enabled advantage. Even in a Big Data world, the emphasis is on deriving insight from the numbers. Few companies smartly pursue the knowledge and wisdom levels of the hierarchy. Harnessing knowledge and wisdom virtually ensures sustained success.

Finally, ask yourself, "Why do most firms find it difficult to tap into knowledge and wisdom?" The answer comes back to management, budgets, and investment patterns. Senior leaders profess that they want to invest in higher-order capabilities because they recognize their importance and they want to win. But, a close look at investment patterns shows that budgets are heavily weighed toward technologies that capture data and attempt to turn it into information. Similarly, despite persistent pleas that "people are our most important asset," even a cursory look at budgets shows that most firms invest far more in IT than in developing people. This pattern drives three counterproductive results. First, a talent deficit arises. People don't possess the skills to collaborate and innovate—key skills at the knowledge and wisdom levels of the hierarchy. Second, people lack the confidence to share their best ideas—essential ingredients in knowledge and wisdom creation. Third, people view IT as a threat to the way they do business. Radically reimagined processes represent too much risk! The bottom line: Budgets and investment patterns consign most companies to compete on data and information. They cannot bring people and technology together to change the competitive rules through product and process innovation.

Following a Proven Path

It is no secret that companies struggle with IT, often failing to achieve targeted benefits. The fear/fantasy adoption drivers make it easy to get caught up in an IT fad and fall susceptible to common traps. For instance, the Big Data craze has spawned the following five blunders:[36]

- **Data for data's sake**—Experts say that many companies forget the fundamental question, "What are we trying to accomplish with this project?" This is the "essential IT is the solution" fallacy that has plagued systems implementations for years.

- **Talent gap**—Technology is a great enabler, but people are needed who can make the most of any technology investment. As for Big Data, data scientists are in short supply.

- **Too much data**—Data collection is no longer the challenge it was before the new millennium. Today's challenge is to define actual data needs. Managers often take the risk-averse approach of saying, "We don't need this now, but we might in the future." Yet, they do not organize the data in a way that makes it easy to access and analyze it, raising the costs of using data to drive insight and better decision making.

- **Turf conflict**—Big Data can drive unique and beneficial decisions, but execution often crosses organizational boundaries. Ownership of projects and execution plans can waylay Big Data initiatives.

- **Overreaching**—A tendency exists to prove a new technology by "showing what it can do." Big projects, however, are often hard to execute—and they often involve big downside risk. Starting small and building quick wins can be a better approach to building momentum.

Following a proven path to IT enablement—such as the one shown in Figure 5-5—can help you avoid such blunders as you maximize your return on IT investments:

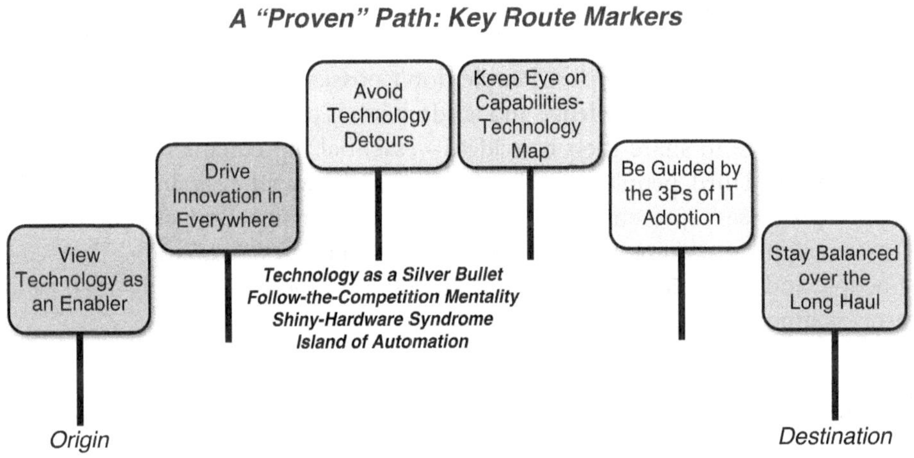

A "Proven" Path: Key Route Markers

```
                    Avoid        Keep Eye on
                    Technology   Capabilities-
                    Detours      Technology
                                 Map
            Drive                              Be Guided by
            Innovation in                      the 3Ps of IT
            Everywhere                         Adoption
View                                                          Stay Balanced
Technology as    Technology as a Silver Bullet                over the
an Enabler       Follow-the-Competition Mentality             Long Haul
                 Shiny-Hardware Syndrome
                 Island of Automation

Origin                                                        Destination
```

Figure 5-5 Markers along the route to IT enablement

- **Route marker 1: Technology as an enabler**—IT is never a panacea. It is always an enabler—of better decision making, improved processes, stronger relationships, and more effective collaboration. One senior manager shared the following warning: "There are no silver bullets, but there are plenty of people willing to sell you one." By staying focused on IT's enabling role, you will not exclude vital organizational issues like trust and willingness from the decision and design processes. High-performing implementations are balanced. The right enabling technologies are acquired and the right culture of people inclusion is encouraged. Winning IT-enabled capabilities result.

- **Route marker 2: Drive innovation**—Most companies find it easier to turbocharge an existing process than to use IT to change the way work is done. Seeking supply chain efficiencies is also easier than improving relationship dynamics. Unfortunately, when IT adoption focuses on efficiency, two counterproductive outcomes result. First, efficiency-driven implementations often alienate the people responsible for implementation and daily use, instilling fear and resistance. As relationships sour, the willingness to share the types of information that can lead to improved operations dissipates. Second, opportunities for meaningful process redesign and supply chain collaboration are overlooked. When these two outcomes emerge, IT implementations deliver mixed results.

- **Route marker 3: Avoid technology traps**—When IT is adopted for the wrong reasons (see route markers #1 and #2 above), companies fall into common traps. Like the blunders that are currently plaguing Big Data, the following generalized traps can undermine your ability to use IT to enhance your firm's capabilities:

  - *Technology as the silver bullet*—Many companies find it easier to buy IT than to fix cultures or redesign underperforming processes that impede competitiveness.

  - *Follow-the-competition mentality*—Trying to avoid being outgunned in an IT arms race makes it easy to play defense and invest in the technologies "everyone" is buying. When everyone else is investing, it can be difficult to just say no.

  - *Shiny-hardware syndrome*—Sometimes the need for an IT investment is evident. However, rather than buying the basic technology to get the job done, managers get caught up in the quest for the latest "shiny hardware (or software)" and overspend in the process.

- *Island of automation*—Isolated investments can solve local problems even as they create a new bottleneck. You need to evaluate IT decisions using a holistic perspective that evaluates the entire system's effect if you want to improve overall performance.

Many managers stumble into one of these traps thinking they are making a powerful investment only to find they have entered into a painful, costly implementation that delivers minimal returns.

- **Route marker 4: Keep your eye on the capabilities-technology map**—A surprising trait among companies that leverage IT for competitive advantage is that they often use technologies others consider as "old." They stick with what they have until they document a need to invest. Disciplined investing is the source of their success. Specifically, investments always map to desired strategic capabilities. For instance, at one Fortune 100 company, the wall of a conference room displays a capabilities-priority chart. The capabilities the company has strategically chosen to build are listed across the top row of the chart. Listed down the left side are all proposed investments. Investments that do not support specific capabilities are not made—period! This disciplined approach changes the IT investment culture. Visibility is created and needs communicated. Because people know why each technology is being adopted and how the technology will improve strategic positioning and operational performance, technophobia is mitigated. Finally, technology traps are avoided.

- **Route marker 5: Be guided by the three Ps of technology adoption**—IT adoption disrupts organizational culture. Needless to say, disruptions are uncomfortable. People fear and resist rather than embrace them. Companies often handle the implementation process awkwardly (they miss route marker #4). To avoid this awkwardness, keep the three Ps of IT adoption in focus:

    - *Process*—IT enablement is all about improving, perhaps revolutionizing, process capabilities. This reality presents a two-way interaction that can be difficult to balance. IT enables process reimagination; however, desired process capabilities must define the IT need—not the other way around. Let the reimagined process guide IT investments.

    - *Performance measurement*—Tom Peters said, "What gets measured, gets done."[37] If your process changes dramatically, you probably need to change your measures. If you don't, you will send your people conflicting messages. Process change communicates one message. Old measures sound a different message. People will listen to the measurement message. Measures need to support the behaviors and skills needed to make the new process work.

- *People*—Process redesign affects people's jobs—and thus their livelihood. You can't afford to forget this fact. You need to make sure that IT "enables" positive changes for the people who will be responsible for executing the new process. For IT adoption to provide a payoff, people need to embrace it. If people are reluctant to use a new process or to share critical information, your IT investment will yield disappointing results. Recognizing this, one Fortune 500 company adopted the catchphrase, "technology is an enabler, people are the bridge or the barrier."

- **Route marker 6: Stay balanced over the long run**—Beginning the IT-enablement journey with the right focus (route markers #1 and #2) is hard. Staying focused during the duration of what can feel like unending implementations (route markers #4 and #5) is harder. Establishing a culture that stays balanced is the hardest step of all. Early successes build momentum. They can also lead to complacency. The discipline you showed earlier may diminish. You might even think you have mastered the slippery slope of technology. However, you need to keep investing in (1) trust; (2) open and honest communication; and (3) an infrastructure of aligned goals, measurement, and training. Culture and structure must support IT enablement if you want to achieve breakthrough performance.

To summarize, done well, IT enablement can help you improve process, relational, and competitive performance. However, as an organizational capability, IT enablement requires holistic decision-making and capabilities-based investment—in people and in change. With the big picture in view, you need to be familiar with the nuts-and-bolts systems that affect customer relationship management and order fulfillment.

Pieces of the IT-Enablement Puzzle

To keep customers happy, you need to build the right service system (occasionally called the service ecosystem). Service systems are defined in a variety of ways, including the following:

- **Customer contact**—The amount of direct contact a customer has with the service system during the creation of the service varies greatly depending on product service. Going to dinner and out to a movie requires high levels of contact. Satisfaction is all about the experience. Streaming a movie on Netflix to watch at home is a low-contact alternative.

- **Customer involvement**—Contact and involvement are different phenomena. Involvement refers to the level of influence the customer has on the service delivery process. Going to a movie is high contact, but low involvement. Going online to design your custom Mini Cooper raises your involvement dramatically. Thus, whereas technology limits involvement in the movie-streaming example, it enhances involvement in the design of a car. Some people find designing a Mini Cooper almost as much fun as driving the car.

- **Capital intensity**—The amount of money invested in infrastructure and technology systems to meet customer needs determines a service system's level of capital intensity. Capital intensity influences not just contact and involvement, but also satisfaction. Amazon.com's entire business model is built around employing capital in the form of technology (the Internet, customer relationship management software, and order processing systems) and infrastructure (a large and growing geographically distributed network of fulfillment centers) to change consumer behavior. Increasingly, capital intensity is defined by technological intensity. This is true for two reasons. First, the sheer cost of technology systems means that technology expenses often dwarf other capital investments. Second, today's technology has a visible and pervasive influence on how customers interact with the service system (i.e., contact and involvement) and, as a result, how they perceive the quality of their service experience.

As Chapter 4, "Configuring the Network for Successful Fulfillment," illustrated, putting in place the right infrastructure to define and support your service system is like putting together a massive jigsaw puzzle. In fact, Chapter 4 suggested that the puzzle pieces are constantly moving and hinted that you might be blindfolded—at least if you don't take special pains to create visibility through systems thinking. The critical point in this chapter is that technology—especially information technology—represents a vital component of a winning service system. Your technology options are vast—and growing. Table 5-1 identifies and describes some of the IT-enabling tools that companies use to track orders, optimize resource utilization, and make customer-pleasing decisions. You might consider each of these IT tools a piece of your service system puzzle. Two of these tools are particularly important to the functionality and performance of your service system: (1) customer relationship management systems and (2) order processing systems.

Table 5-1 IT-Enabling Tools

	Definition	Brief History	Status	Where It Is Used
RFID	Wireless noncontact use of radio-frequency fields to transfer data for the purpose of automatically identifying and tracking tags attached to objects. RFID tags can be used to track and manage inventory, assets, people, and so on.	1945—An espionage tool that retransmitted incident radio waves was introduced. World War II—An IFF transponder was used to identify aircraft. 1973—First passive radio transponder with memory was introduced. 1983—First RFID patent filed.	Advantage—Unlike bar codes, RFID tags do not have to be visible. Disadvantage—RFID tags cannot be used with products that contain water. The current cost is as little as $.05. The value of the RFID market in 2012 is $7.46 billion.[38]	Automotive—Track assembly line production. Pharmaceuticals—Track through warehouse. Livestock—Used for identification of animals. Memorabilia and art—Track to verify authenticity. Travel—Track airport baggage.
Bar codes	An optical machine-readable representation of data relating to the object to which it is attached. Used primarily to track objects through distribution and to read prices at retail.	1949—Patent for linear and bull's-eye printing filed. 1967—The railroad industry selected KarTrak as a standard.[39] 1972—The first retail use of bar codes scanned the price of chewing gum at a Kroger store.[40]	Bar codes such as the UPC are a universal element in the shopping experience. Bar codes allow for the organization of large amounts of data.	Retail—Most items have a UPC bar code. Healthcare—Use bar codes for medication management. Logistics—Track rental cars, airline luggage, mail, and parcels. Ticketing—Use bar codes for ticketing at sports arenas, cinemas, and fairgrounds.

	Definition	Brief History	Status	Where It Is Used
WMS	A warehouse management system (WMS) that controls the movement and storage of materials within a warehouse and processes the associated transactions, including shipping, receiving, putaway, and picking.	Warehouse management was a manual process and everything was done with paper and pencil. In the late twentieth century, WMS employed technology such as bar codes, radio frequency, and computers to track movement and storage of inventory within the warehouse setting.[41]	Current objectives are to provide a set of computerized procedures for management of warehouse inventory with the goal of minimizing cost and fulfillment times.	Utilize automatic identification and data-capture technology such as bar code scanners, mobile computers, wireless LANs, and, potentially, RFID to monitor the flow of products.
Pick to light	A system in which each pick location is connected to lights and LED displays; software turns on the light where the next pick should be and indicates the quantity to pick.	First PTL systems used floppy disks to get data to the line; most workers still carried paper lists. Today, PTL systems are integrated into the WMS system.[42]	Productivity improvements from PTL systems generally range from 25 percent to 50 percent in as few as 12 months.[43]	Warehouses—Increase productivity and accuracy. Distribution centers—Use high-speed sort capabilities. Order fulfillment—Use real-time sorting of orders and directed replenishment. Use reverse logistics and returns processes.
TMS	A transportation management system that usually sits between an ERP and warehouse/distribution module, where both inbound and outbound orders are evaluated, and offers various suggested routing solutions.	1991—"Blue screens" were only at the most progressive organizations. 1996—Movement toward client/server began, while broader adoption of TMS continued. 2001—Internet-based systems were launched. 2006—Movement toward more integrated intelligence began.[44]	About 39 percent of firms utilize a TMS with 40 percent achieving 90 percent on-time delivery.[45] Recently, diverse types of licensing allow shippers who otherwise would not be able to afford sophisticated software the opportunity to utilize TMS to better manage vital functions.	Manage motor carrier, rail, air, and maritime transport. Use TMS for real-time transportation tracking. Use TMS for vehicle load and route optimization. Use TMS for shipment batching orders.

	Definition	Brief History	Status	Where It Is Used
CRM	Technology that manages a company's interactions with customers to organize, automate, and synchronize sales, marketing, customer service, and technical support.	1980—Businesses began using databases to track existing and potential customers. 1990—Companies began to use CRM to support customer loyalty programs. 2000—CRM software evolved to support company-specific needs, including customer profitability analysis.[46]	There is a current shift from *push CRM* toward a *customer transparency* (CT) model, due to the increased proliferation of channels, devices, and social media.[47] Vendor relationship management (VRM) is another development that provides tools and services that allow customers to manage individual relationships with vendors.	Marketing—Measure campaigns over multiple channels, such as email, search, social media, telephone, and direct mail. Appointments—Integrate emails, documents, jobs, faxes, and scheduling for individual accounts. Small business—Manage events and track relationships. Social media—Make use of social media to build customer relationships.
BI	An application software designed to retrieve, analyze, and report data for business intelligence (BI). The tools generally read data that has been previously stored in a data warehouse or data mart.	1980—Data warehouses were developed so data could be organized and come from multiple locations. 1990—BI vendors started to emerge to provide more access across multiple locations to report and analyze data. 2000—Data marts evolved to organize huge data warehouses into usable data.	While there are plenty of BI start-ups, longtime vendors control the market. Companies spend about $79 billion a year on BI software and services.[48] The key driver in BI is the need to find meaningful information in Big Data and to be able to disseminate that information.[49]	Measurement—Hierarchy of performance metrics inform business leaders about progress toward goals. Analytics—Businesses use this quantitative process to arrive at optimal decisions and perform knowledge discovery. Collaboration—BI helps different companies to work together through data sharing and electronic data interchange (EDI).

	Definition	Brief History	Status	Where It Is Used
EDI	A document standard that, when implemented, acts as a common interface between two or more computer applications in terms of understanding the document transmitted.	1980—EDI standards were first implemented in the automotive industry. 2002—The Internet Engineering Task Force (IETF) published RFC 3355, offering a standardized, secure method of EDI data via email. 2005—An IETF working group ratified RFC 4130 for MIME-based HTTP EDINT transfers.[50]	Data—Every industry uses different documents that have to be adapted to EDI documents. Standards—Different parts of the world may use ANSI X12 standards versus EDIFACT. Communications—Different industries have adopted different protocols for sharing information. Formats—EDI can include many formats for documents, which can vary from industry to industry.	Healthcare—HIPAA requirements establish EDI as a requirement in healthcare. Automotive—EDI streamlines operations for automotive OEMs and Tier 1 suppliers. Retail—EDI lowers prices through streamlined processes. Manufacturers—Benefit from EDI in all product categories. Suppliers—Businesses look to utilize EDI across their entire supply chain.

Customer Relationship Management Systems

Customer relationship management (CRM) systems enable companies to get to know their customers, their wants, and their buying habits. Specifically, CRM can help you create accurate customer profiles—a form of very sophisticated customer segmentation. Consider the following description of how data analytics enables segmentation from *The Wall Street Journal*:

> You may not know a company called [x+1] Inc., but it may well know a lot about you.
>
> From a single click on a web site, [x+1] correctly identified Carrie Isaac as a young Colorado Springs parent who lives on about $50,000 a year, shops at Walmart and rents kids' videos. The company deduced that Paul Boulifard, a Nashville architect, is childless, likes to travel and buys used cars. And [x+1] determined that Thomas Burney, a Colorado building contractor, is a skier with a college degree and looks like he has good credit.
>
> The company didn't get every detail correct. But its ability to make snap assessments of individuals is accurate enough that Capital One Financial Corp. uses [x+1]'s calculations to instantly decide which credit cards to show first-time visitors to its web site.
>
> *In short: Web sites are gaining the ability to decide whether or not you'd be a good customer, before you tell them a single thing about yourself.*[51]

When you possess intimate insight into your customers' buying habits, you can develop the right products, locate them where customers want to buy them when customers want to buy them, and provide the right service experience—all at lower costs. This is the goal of outstanding order fulfillment and customer service. CRM systems consist of four primary technology components: data-capture technology, data storage, data analytics, and information display systems:

- **Data-capture technology**—Although each of the four components can be used to one-up rivals, without good data-capture technology, the remaining components are relegated to nice-to-have-but-not-very-useful toys. That is, your analysis is held captive by the lack of data. You need to capture accurate data to gain insight into the habits and needs of your customers. In this arena, the old *GIGO* computer adage holds true—garbage in, garbage out.

    E-commerce entities, thus, have a distinct advantage over their bricks-and-mortar-only rivals. You already know why. Well-placed cookies allow Internet retailers to track every "click" you make while browsing their Web sites. They even know how long you stay on each page. Amazon set the early standard, creating customer profiles and using data from your past visits to suggest additional titles that might be of interest to you. Victoria's Secret set a little sexier tone, using online click data to redesign the Web storefront and optimize sales. In one case, a lace-up gown wasn't selling on the page reserved for gowns, but was a fast mover on Victoria Secret's Top 10 page. Sales increased when the gown was displayed more conspicuously in the gown section. Even better, sales of other gowns also increased.[52] The bigger challenge for Internet entities, until recently, was to navigate the vast oceans of data to figure out which clicks really explained buying behavior.

Bricks-and-mortar stores have moved aggressively to close the data-capture gap. Their weapon of choice is the loyalty card. Whenever you use your card, transaction details are captured and stored in a customer database, making it easy for your favorite stores to get to know your buying habits. Walmart has never embraced loyalty cards, preferring to offer EDLP pricing to all of its customers. However, not wanting to be left out of the data game, Walmart is piloting self-scan technology that not only lets customers avoid the checkout line, but also provides Walmart insight into both what customers buy and how they buy it.[53] Other stores are using video footage from security cameras to track customer shopping patterns.[54] All of this data can now be mixed and matched via Big Data techniques to help managers decide exactly how to meet customer needs more completely.

- **Data storage**—The key point to know about data storage is that it is expanding exponentially. For example, in 2008, Li & Fung processed 100 gigabytes per day. By 2010, the data flow had increased tenfold.[55] In 2009, the Internet stored 500 exabytes. In 2012, the Web kept 2.5 zettabytes of data "secure." Facebook alone collected 15 million terabytes—each and every day. Analysts expect the figure will reach 8 zettabytes by 2015.[56] If you don't speak the language of bytes, the basic translation follows (the typical laptop has 200–500 gigabytes of hard-disk storage):

    A gigabyte is 1,024 megabytes.

    A terabyte is 1,024 gigabytes.

    A petabyte is 1,024 terabytes.

    An exabyte is 1,024 petabytes.

    A zettabyte is 1,024 exabytes.

This expansion of storage capacity has given rise to Big Data and is changing the way companies view and manage data. For example, companies now rent data storage space in the cloud rather than buying server capacity. They also access software from the cloud, treating software acquisition as a service; in fact, the practice is called software as a service (SaaS).

- **Data analytics**—Now that data is so easy to collect (think clicks, bar codes, RFID, sensors, and loyalty cards) and so inexpensive to store, analytical capability is now the bottleneck in translating data to capability. Data analytics has, thus, become a source of competitive advantage. Who would have thought that a book titled, *Competing on Analytics: The New Science of Winning*, would become a best seller?[57] The data scientists responsible for developing the analytic programs that crunch the numbers are in such demand that they earn salaries ranging from $150,000 to $500,000 a year.[58] The analysis is so sophisticated that many companies outsource the analytics. For example, Kroger relies on the British firm Dunnhumby to dig into the data. Rodney McMullen, vice chairman at Kroger, says, "Our partnership with Dunnhumby is one of our key competitive advantages.... In our competitive industry, remaining relevant to our customers is critical."[59] So, what are companies learning via analytics? Simply stated, they are learning how to increase profits and customer satisfaction simultaneously. Consider how retailers are using the data to change the way they do business:

  - *Targeted promotions*—Kroger reduced its reliance on circulars, preferring to reach *loyal* customers through targeted coupon packages.[60] Tesco actually began mailing beer coupons to shoppers who bought diapers. Why? Analysis revealed that new fathers who are stuck at home tending the baby drank more beer. Karen Masek, a Tesco shopper, commented, "They definitely know your shopping habits. They've never sent me anything totally off the mark."[61] By reaching out to loyal customers with targeted promotions, Kroger and other retailers reduce "cherry picking" by customers who only buy items that are on sale. On the Web, analytics make it possible for retailers to tailor their Web sites to show different visitors different products—and different prices.[62]

  - *Product selection*—Kroger began stocking hard-to-find brands in order to increase loyalty among high-margin demographics. Because it is a slow seller, Alpen breakfast cereal does not create the flow through or the margin to earn a place on the shelf. However, health-conscious customers who purchase Alpen tend to fill their carts with a range of high-margin products.[63]

  - *New product development*—Kroger introduced new products. For example, three-quarter-gallon milk cartons fit the needs of today's smaller families better than traditional half or whole gallon options.[64] Tesco developed its highly popular "Tesco Finest" line when it discovered what customers weren't buying at Tesco: wine, cheese, and fruit.[65]

  - *Product display*—Kroger developed new merchandizing plans. Coffee, which was always placed on shelves by brand, is now placed into caffeinated and decaf sections.[66]

- **Pricing**—Tesco thwarted tough, low-price rivals like Walmart's Asda by singling out price-sensitive customers who might switch loyalty to save a little money. By lowering prices on a set of 300 items these shoppers bought regularly, Tesco kept them from visiting Asda to comparison shop.[67]

- **Data display**—Before data becomes truly useful, you need to get the results of your analysis into the hands of decision makers in a form that they can make sense of and use in decision making. Otherwise, all of your investments are for naught. The key is to make relationships that are hidden in the data visible to the decision maker. Consider the following example:

  Two and a half years ago, Pek Lum, a biologist by training, was looking for a cancer cure when she stumbled upon clues that were hiding in plain sight.

  Poring once more over a 12-year-old set of data on breast-cancer tumors, Dr. Lum saw correlations between the disease and patients' outcomes that she and her fellow researchers had never noticed before—correlations that eventually could lead to more-effective breast-cancer treatments.

  Nothing about the data had changed. Except the way it looked. Dr. Lum's new view came courtesy of software that uses topology, a branch of math that compresses relationships in complex data into shapes researchers can manipulate and probe: in this case, a Y, like a two-eared worm.[68]

Of course, decision-making visibility does not always require advanced, esoteric math. Classic examples of how companies have displayed analytical results to enhance personalization and profitability follow:

- The Limited added a personal flair to its customer profiling, encouraging salespeople to enter comments on customer preferences and past purchases into its CRM records. During the next visit to the store, the salesperson can personalize the interaction, asking questions like, "How do you like the red sweater you bought last month?" or "Are you enjoying your art classes?"[69]

- First Union Corp. was one of the first financial-service providers to employ CRM to segment customers. When a customer called for assistance, a colored square appeared on the service representative's screen, indicating the level of service that could be provided. Green enabled the waiving of fees; red meant polite service but no special help.[70]

- Continental Airlines armed its employees with a system that let them see a passenger's flying history—and an estimation of the passenger's value to the airline. Each employee could add comments about passenger preferences to help improve the next service experiences. One result of higher levels of customer intimacy: The number of passengers flying on higher-cost, unrestricted fares increased by 25 percent.[71]

As fashionable as CRM systems have become, you should temper—at least for now—your enthusiasm for data analytics-driven customer relationship management. At least, you should understand CRM's dark side. Over a decade ago, *BusinessWeek* ran a cover story titled, "Why Service Stinks." The tagline was tantalizing: "Companies know just how good a customer you are—and unless you're a high roller, they would rather lose you than take the time to fix your problem." The reality is that current analytical tools provide great insight into correlation, but not causality. Some people do not see this as a problem. For example, in their book, *Big Data: A Revolution That Will Transform How We Live, Work, and Think,* Viktor Mayer-Schönberger and Kenneth Neil Cukier write: "Society will need to shed some of its obsession for causality in exchange for simple correlations: not knowing why but only what. This overturns centuries of established practices and challenges our most basic understanding of how to make decisions and comprehend reality."

However, blind trust in the data can be dangerous when dealing with customers. Past buying behavior does not always predict future behavior. In today's mobile society—exacerbated by a turbulent economy—situations and lifestyles change quickly, transforming today's nuisance customer to tomorrow's preferred customer (or vice versa). If a customer perceives that she has a red square by her name and is not valued by your company, she may take her business elsewhere—forever. Such a decision could cost your company substantial profit over a 50-year spending spree. By extension, low levels of activity may reflect more on a company's service offering than on the quality or profitability of a customer. Perhaps the customer had a bad service experience with the company. Fixing the problem may salvage this—and many similar—customer relationships. The bottom line: Correlation and causality each have their place in the design of winning fulfillment and service systems. Well-designed CRM systems can help you treat every customer contact as an opportunity to create a profitable customer.

Order Processing Systems

Order processing systems don't capture the imagination or garner the attention of CRM systems, but they are critically important to building a winning order fulfillment capability. Order processing initiates order fulfillment (see Figure 5-6, Panel A). The order processing system captures order information, stores it in a central database, and transmits it to decision makers in logistics and accounting so that the picking/shipping and billing processes can be efficiently executed (see Figure 5-6, Panel B). A well-designed order processing system enables you to keep track of order data, maintain visibility into the physical status and positioning of individual orders, and monitor inventory levels for each step of the fulfillment process. The information captured by your order processing system plays a key role in developing future forecasts and marketing plans.

Panel A: Order Processing's Place in Order Fulfillment

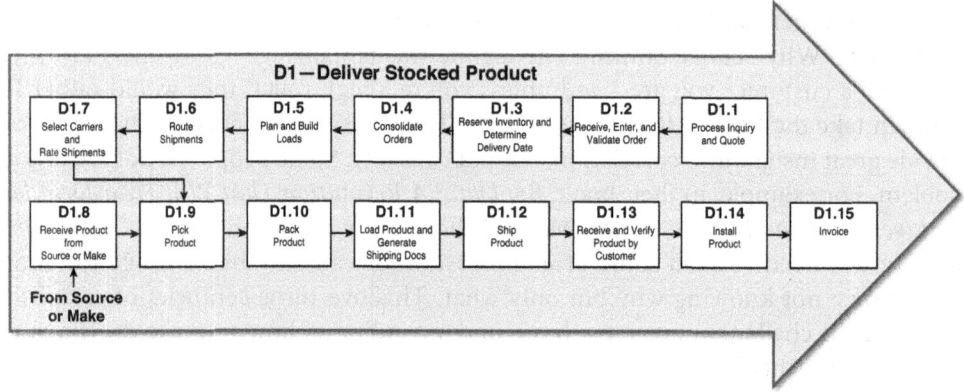

Panel B: The Basic Architecture of an Order Processing System

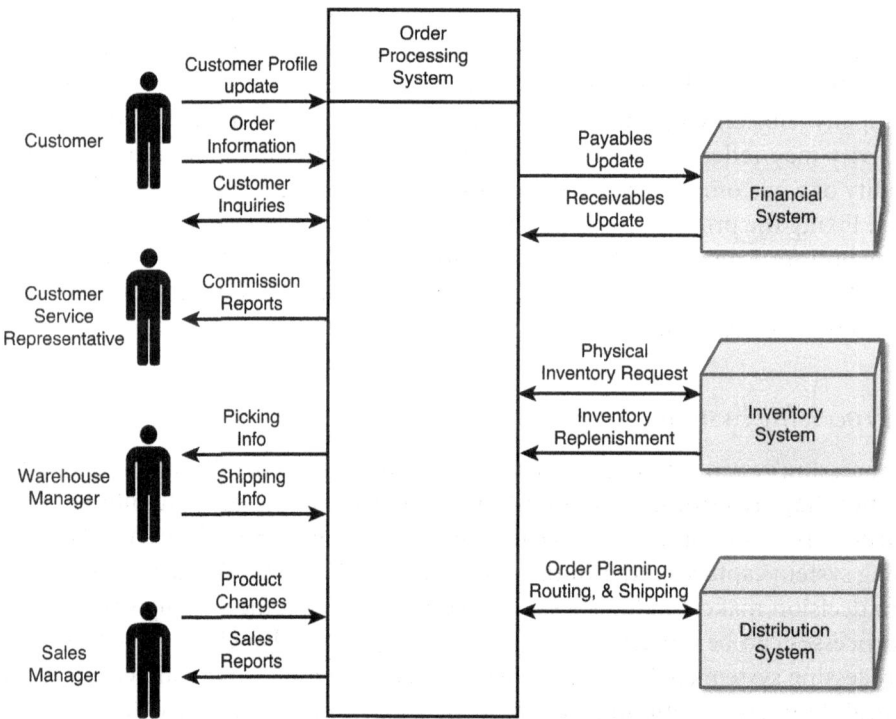

Figure 5-6 The architecture of an order processing system

160 THE DEFINITIVE GUIDE TO ORDER FULFILLMENT AND CUSTOMER SERVICE

Historically, order processing has been highly labor intensive. A customer service representative would take orders by phone, by fax, or in person and enter them into the company's order system. Not only was the order entry process time consuming, but it also introduced numerous errors into the task of filling an order. It was not uncommon for a manual process to take 48–72 hours to get an order into the system. Moreover, poor order entry (e.g., transcription errors) was a leading cause of inaccurately filled orders—a major source of customer angst.

Although some companies still employ a high degree of manual labor, most use an automated order entry system that allows customers to enter orders directly via the Internet or electronic data interchange (EDI). Automated order processing frees up most of the two to three days previously needed for order entry. This time can be used to offer shorter delivery cycles or improve operating efficiencies. For example, the extra time can facilitate order consolidation, improved vehicle routing, and shipping efficiencies (lower-cost options or fuller trucks). Further, an extra 72 hours can allow improved production scheduling or alternative shipping such as direct shipping of consolidated orders from the factory—a move that can eliminate materials handling at the DC.

To best determine how improved order processing capabilities can enable you to offer a more competitive value proposition to the customer, you need to run the numbers to evaluate all of the relevant system tradeoffs. Specifically, automated order processing systems can be costly and difficult to implement and maintain, requiring high-skill IT personnel. These costs limit the adoption of advanced systems among very small companies. However, the benefits can be substantial (see Table 5-2) and can be divided into two categories: service and asset utilization. Advanced order management systems take time out of the order fulfillment process even as they increase order fill accuracies. When customers receive exactly what they order fast, they are satisfied. As for asset utilization, good—accurate and timely—data is needed to plan operations, including picking, packing, and shipping. Getting the right information into the hands of the people doing the work when they need it makes their job easier, increasing efficiencies. For example, the order processing system feeds into the warehouse management system, which establishes picking priorities and routines. The right items get picked for each order the first time.

Table 5-2 Benefits of an Advanced Order Management System

Customer Service	
Time	Automated systems reduce the initial time required to initiate the rest of the order fulfillment process. More accurate order information reduces rework. The automated system shares information to speed up other order fulfillment activities like picking, packing, vehicle loading, and routing. The result: faster cycles and more responsive delivery.
Accuracy	Data only needs to be entered one time—typically by the customer during the order placement. Delivering the right orders on time is vital to meeting customer expectations.

Asset Utilization		
Inventory	Better information accuracy reduces the need for just-in-case and redundant inventories. Higher-quality, more-timely information also improves the reliability of forecasts and the efficacy of operating plans. Shorter, less-variable delivery cycles enable customers to reduce their inventories.	
Warehousing	The planning enabled by a well-designed order processing system—in conjunction with a warehouse management system—enables more effective order consolidation, more accurate and efficient picking, and better utilization of warehouse resources (people and equipment).	
Transportation	The planning enabled by a well-designed order processing system—in conjunction with a transportation management system—enables more effective shipment consolidation and better routing even as it reduces the need for expediting.	
Production	Better information regarding customer orders and product demand improves forecast accuracy, leading to better short-term production scheduling. Greater efficiencies and shorter cycles throughout the order fulfillment process also provide slack, reducing the need for expedited production.	

Conclusion

The information revolution has not only enabled modern supply chain management, but it has also created the demand for world-class supply chain execution. Consider for a moment television's influence on global demand patterns. Over the past 30 years, shows like *Dallas* (1980s), *Baywatch* (1990s), and *Friends* (1990s-2000s) have become globally syndicated hits. What does this mean? Consumers worldwide know how other people around the world live. With this increased awareness, they want the cool iPhone or the latest edition of a Mercedes sedan. Thus, companies need to build the global production and distribution capabilities needed to efficiently utilize worldwide resources to meet global customer needs. If you can't deliver what the customer wants, someone else will—creating new rivals. Companies like BlackBerry and Nokia did not fully grasp this reality until it was too late. Andy Grove summarized this phenomenon:

> The new environment dictates two rules: First, everything happens faster; second, anything that can be done will be done, if not by you, then by someone else, somewhere. *Let there be no misunderstanding: These changes lead to a less kind, less gentle, and less predictable workplace.*[72]

Even as TV has changed customer aspirations, digitization has changed value-creation capabilities. The Internet enables instantaneous global communication and global order capture. Advanced systems—from CRM to ERP to WMS—enable companies to really get into the lives of customers and efficiently plan operations to create the value customers demand. These are powerful, competitive weapons. Because you know the competition is buying them, you feel compelled to as well. Unfortunately, buying IT and using IT for competitive advantage are not synonymous. Remember, there are no silver bullets. To use IT as an enabler of better order fulfillment, higher levels of customer satisfaction, and improved competitive advantage, you need the following:

- **Deep understanding**—What are you trying to accomplish? If you don't know what specific capabilities you are trying to build, you will likely invest in the wrong IT and then implement it inappropriately.

- **Good data**—Before you can use data to build information, knowledge, or wisdom capabilities, you need accurate and relevant data—and you need it to be fresh. In today's fast-paced world, data comes with an expiration date.

- **Skilled people**—Getting data to reveal its secrets requires a new type of employee—the data scientist. Turning knowledge into insight and then into customer-pleasing capabilities requires experienced managers who are comfortable with both technology and change.

- **Emotional fortitude**—In their book, *Competing on Analytics: The New Science of Winning*, Thomas Davenport and Jeanne Harris emphasize the following key point: You need managers who have the guts to stick to decisions based on the analytics—even if they don't fully understand why the world works the way it does.[73] Because "everything happens faster," you seek correlation-driven, first-mover advantage even as you strive for causation-driven understanding.

As you weigh the costs inherent in developing these skills, you might be wondering, "We've done OK so far. Do we really need to invest in these skills?" The answer is a resounding, "Yes." The IT revolution is really just getting started. Near-term technology advances like additive manufacturing are going to change supply chain design. Never before has it been so easy to design and manufacture a complex product as it is today. Consider TV comedian Jay Leno. As an avid car collector, one of his greatest annoyances has been to find the parts to keep his 200-car collection—which includes a 1906 Stanley Steamer—in driving condition. Today, he can make one-of-a-kind-parts using a NextEngine scanner and a Stratasys 3-D printer. Mr. Leno notes, "The days of going to a junkyard and trying to find an auto part that says Packard or Franklin on it are over. We can make almost anything we need right here in the shop ourselves." Mr. Leno explains the process of creating a replacement part:

> We took the worn piece and copied it with a scanner that can measure about 50,000 points per second.... That created a digital file or image of the part, which we can modify in the computer if there are imperfections or defects in the part being scanned. Then you feed that data into the 3-D printer, and, presto, you have a mold that will allow you to cast a brand new part.... The new tools have really changed the game. We can be pretty much self-sufficient here.[74]

Not everyone can afford their own 3-D equipment, but third-party design shops[75] and logistics service providers like UPS[76] are investing in these technologies to bring them within your reach. Additive technologies are poorly suited for some highly specialized parts (e.g., those that are exposed to extreme conditions) and cannot match mass production's efficiency. However, imagine the potential for Caterpillar to redesign its spare parts' supply chain or for the military to support deployed aircraft. The cost/availability equation is powerful and it will change order fulfillment models. The additive manufacturing and the Internet are not quite Star Trek's replicator and transporter technology, but they do raise the question, "What if?" and they dictate that you continue to scan, crunch the numbers, and reimagine your fulfillment capabilities.

Endnotes

1. Westerman, G. 2012. "IT Is from Venus, Non-IT Is from Mars." *The Wall Street Journal,* April 2.
2. Koch, C. 2002. "Supply Chain: Hershey's Bittersweet Lesson." *CIO,* November 15. Retrieved September 25, 2013, from http://www.cio.com/article/31518/Supply_Chain_Hershey_s_Bittersweet_Lesson
3. Roberti, M. 2003. "Analysis: RFID—Walmart's Network Effect." *CIO Insight,* September 15. Retrieved September 25, 2013, from http://www.cioinsight.com/c/a/Trends/Analysis-RFID-WalMarts-Network-Effect/
4. "Getting an Edge from IT." 2009. *The Wall Street Journal,* November 30.
5. Hammer, M. 1990. "Reengineering Work: Don't Automate, Obliterate." *Harvard Business Review* 68(4):104–131.
6. Fawcett et al. 2006. "Managing the Slippery Slope of Technology." *Supply Chain Management Review*, 12(7): 14-20.
7. Fawcett, S., Wallin, C., Allred, C., Fawcett, A., and Magnan, G. 2011. "Information Technology as an Enabler of Supply Chain Collaboration: A Dynamic-Capabilities Perspective." *Journal of Supply Chain Management* 47(1):38–59.
8. Kuglin, F. 1998. *Customer-Centered Supply Chain Management.* New York, NY: AMACOM.
9. Arthur, W. 2003. "Why Tech Is Still the Future." *Fortune,* November 24. Retrieved September 14, 2013, from http://money.cnn.com/magazines/fortune/fortune_archive/2003/11/24/353778/

10. "The Greatest Defunct Web Sites and Dotcom Disasters." 2008. *CNET,* June 5. Retrieved September 14, 2013, from http://web.archive.org/web/20080607211840/ http://crave.cnet.co.uk/0,39029477,49296926-6,00.htm

11. Fawcett, S., Wallin, C., Allred, C., and Magnan, G. 2009. "Supply Chain Information Sharing: Benchmarking a Proven Path." *Benchmarking: An International Journal* 16(2):222–246; {Fawcett, 2007 #3526}

12. Ibid, 11

13. Ibid, 6

14. Fawcett, S. 2000. *The Supply Management Environment: Supply Management's Role Today and in the Future.* Phoenix, AZ: National Association of Purchasing Management.

15. Koten, J. 2013. "A Revolution in the Making." *The Wall Street Journal,* June 10:R1.

16. Hagerty, J. 2013. "How Many Turns in a Screw? Big Data Knows." *The Wall Street Journal,* May 15:A8.

17. Smith, L., Andraski, J., and Fawcett, S. 2011. "Integrated Business Planning: A Roadmap to Linking S&OP and CPFR." *Journal of Business Forecasting* 29(4):4–13.

18. Garry, M. 2006. "With RFID, P&G Improves Launch of Fusion Razors." *Supermarket News,* March 27. Retrieved September 14, 2013, from http://supermarketnews.com/archive/rfid-pg-improves-launch-fusion-razors

19. Schuman, E. 2009. "P&G's Decision to Pull Back from Walmart RFID Trial Quite Understandable." *StorefrontBacktalk,* February 19. Retrieved September 14, 2013, from http://storefrontbacktalk.com/supply-chain/ pgs-decision-to-pull-back-from-Walmart-rfid-trial-quite-understandable/

20. Anderson, G. 2011. "Macy's Moves to Item-Level Tracking Using RFID." *Retailwire,* September 29. Retrieved September 14, 2013, from http://www.retailwire.com/discussion/15539/macys-moves-to-item-level-tracking-using-rfid

21. Berman, D. 2013. "Daddy, What Was a Truck Driver?" *The Wall Street Journal,* July 23.

22. McCarter, M., and Northcraft, G. 2007. "Happy Together?: Insights and Implications of Viewing Managed Supply Chains as a Social Dilemma." *Journal of Operations Management* 25(2):498–511; Fawcett, S., Magnan, G., and McCarter, M. 2008. "Supply Chain Alliances and Social Dilemmas: Bridging the Barriers That Impede Collaboration." *International Journal of Procurement Management* 1(3):318–341.

23. Ibid, 11

24. Hall, G., Rosenthal, J., and Wade, J. 1993. "How to Make Reengineering Really Work." *Harvard Business Review* 71(5):119–131.

25. Hammer, M. 2004. "Deep Change." *Harvard Business Review* 82(4):84–93.

26. Rosenbush, S., and Totty, M. 2013. "How Big Data Is Changing the Whole Equation for Business." *The Wall Street Journal,* March 8.

27. Gage, D. 2013. "The New Shape of Big Data." *The Wall Street Journal,* March 8.

28. Steel, E., and Angwin, J. 2010. "On the Web's Cutting Edge, Anonymity in Name Only." *The Wall Street Journal,* August 4.

29. Ovide, S. 2013. "Big Data, Big Blunders." *The Wall Street Journal,* March 8.

30. Ibid, 6

31. Dayton, B. 2011. Innovationlive: Engaging 3M's Global Employees in Creating an Exciting, Growth-Focused Future. June 16. Retrieved September 16, 2013, from http://www.managementexchange.com/story/innovationlive-engaging-3ms-global-employees-creating-exciting-growth-focused-future

32. Gaberman, I., and Witjes, M. 2011. "Demystifying Corporate Culture." *ATKearney,* Retrieved September 16, 2013, from http://www.germany.atkearney.com/organization-transformation/ideas-insights/article/-/asset_publisher/LCcgOeS4t85g/content/demystifying-corporate-culture/10192; Wallace, G., and Garcia, S. 2011. "HR's Innovative Role in Creating High-Impact Online Collaborative Communities." *Workforce Solutions Review,* Retrieved September 16, 2013, from http://www.philosophyib.com/3/case/Wallace_Garcia_WSR.pdf

33. Useem, J. 2003. "Fortune 500." *Fortune,* April 14:87–90.

34. Chichester, G. C. 2003. "No. 1 Walmart: Letter to Editor." *Fortune*: 30.

35. Fawcett, S., Rhoads, G., and Burnah, P. 2004. "People as the Bridge to Competitiveness: Benchmarking The 'ABCs' of an Empowered Workforce." *Benchmarking an International Journal* 11(4):246–360; Fawcett, S. E., Brau, J. C., Rhoads, G. K., Whitlark, D., and Fawcett, A. M. 2008. "Spirituality and Organizational Culture: Cultivating the ABCs of an Inspiring Workplace." *International Journal of Public Administration* 31:420–438.

36. Ibid, 29

37. Peters, T. 1986. "What Gets Measured Gets Done." *tompeters Column Archives,* April 28. Retrieved September 23, 2013, from http://www.tompeters.com/column/1986/005143.php

38. "New Forecast Released Says RFID Market Will Be $7.46 Billion in 2012." 2012. *RFID World Canada,* June 12. Retrieved September 25, 2013, from http://www.rfidworld.ca/new-forecast-released-says-rfid-market-will-be-7-46-billion-in-2012/917

39. Graham-White, S. 1999. "Do You Know Where Your Boxcar Is?" *Trains* 59(8):48–53.

40. Nelson, B. 1997. *From Punched Cards to Bar Codes: A 200 Year Journey.* Chicago, IL: Helmers Pub. Co.

41. Lucianocunha. 2012. "The Evolution of the Warehouse Management System." *Microsoft Dynamics Community,* June 20. Retrieved September 25, 2013, from http://community.dynamics.com/b/toincreaseblog/archive/2012/06/20/the-evolution-of-the-warehouse-management-system.aspx#.UjuwR4KAGF0

42. Morris, J. 2013. "The Evolution of Intralogistics." *Intralogistics, Material Handling Consulting,* June 6. Retrieved September 25, 2013, from http://www.invata.com/the-evolution-of-intralogistics/

43. "Pick to Light and Put to Light." 2013. *Genco,* Retrieved September 19, 2013, from http://www.genco.com/distribution/pick-to-light-put-to-light.php

44. Computer Science Corporation. 2008. *Transportation Management: A New Landscape* (November): APICS Chicago Chapter.

45. Ibid, 44

46. Koble, M. 2013. "The History of CRM Software." *eHow Tech,* Retrieved September 19, 2013, from http://www.ehow.com/about_6573510_history-crm-software.html

47. DeGregor, D. 2011. *Customer-Transparent Enterprise: Beyond 20th Century CRM.* Henderson, NV: Motivational Press.

48. Robb, D. 2013. "What's New with 5 Big Bi Vendors." *Enterprise Apps Today,* July 23. Retrieved September 19, 2013, from http://www.enterpriseappstoday.com/business-intelligence/whats-new-with-5-big-bi-vendors.html

49. Ibid, 48

50. Kantor, M, Ed. Burrows, J. (1996-04-29). "Electronic Data Interchange (EDI)." National Institute of Standards and Technology. Retrieved September 25, 2013.

51. Steel, E., and Angwin, J. 2010. "On the Web's Cutting Edge, Anonymity in Name Only." *The Wall Street Journal,* August 4.

52. Totty, M. 2003. "How Can E-Tailers Get to Know You Better." *The Wall Street Journal,* October 20:R4.

53. Wohl, J. 2013. "Walmart Adds iPhone Scan-and-Checkout Feature to 12 More Markets." *Reuters,* March 20. Retrieved September 25, 2013, from http://www.reuters.com/article/2013/03/20/us-walmart-checkout-expansion-idUSBRE92J0P020130320

54. Clifford, S., and Hardy, Q. 2013. "Attention, Shoppers: Store Is Tracking Your Cell." *The New York Times,* July 14. Retrieved September 25, 2013, from http://www.nytimes.com/2013/07/15/business/attention-shopper-stores-are-tracking-your-cell.html?pagewanted=all&_r=0

55. "The Data Deluge." 2010. *The Economist,* February 25.

56. Deutscher, M. "When Will the World Reach 8 Zetabytes of Stored Data?" *Silicon Angle,* Retrieved September 18, 2013, from http://siliconangle.com/blog/2012/05/21/when-will-the-world-reach-8-zetabytes-of-stored-data-infographic/di

57. Davenport, T., and Harris, J. 2008. *Competing on Analytics: The New Science of Winning,* Harvard Business School Press: Boston MA.

58. Hadi, M. 2007. "The Importance of Crunching the Numbers." *The Wall Street Journal,* April 18:D12.

59. Rohwedder, C. 2007. "What Makes Tesco, Kroger More Than Just Rivals?" *The Wall Street Journal,* December 24:B1.

60. Ibid, 59

61. Ibid, 51

62. Rohwedder, C. 2006. "No 1 Retailer in Britain Uses 'Clubcard' to Thwart Wal-Mart." *The Wall Street Journal,* June 6:A1.

63. Ibid, 59

64. Ibid, 59

65. Ibid, 62

66. Ibid, 59

67. Ibid, 62

68. Ibid, 27

69. Blackwell, R. 1997. *From Mind to Market: Reinventing the Retail Supply Chain.* New York: Harper Business.

70. Brady, D. 2000. "Why Service Stinks." *BusinessWeek* (October 23).

71. Tyndall, G. 1998. *Supercharging Supply Chains.* New York: John Wiley & Sons, Inc.

72. O'Grady, B. 2012. Tesco, Sony, Nokia: "Why Companies Come a Cropper and What to Learn." *WordPress.com,* April 15. Retrieved September 25, 2013, from http://brianogrady.wordpress.com/2012/04/15/tesco-sony-nokia-why-companies-come-a-cropper-and-what-to-learn/

73. Ibid, 57

74. Koten, J. 2013. "Who Says Jay Leno Isn't Cutting Edge?" *The Wall Street Journal,* June 10.

75. Fowler, G. 2013. "Build a Better Mousetrap—Fast." *The Wall Street Journal,* June 10.

76. Diakov, D. 2013. "UPS May Have Hit Pay Dirt with 3D Printing." *Forbes,* Retrieved September 20, 2013, from http://www.forbes.com/sites/rakesh-sharma/2013/08/19/ups-may-have-hit-pay-dirt-with-3d-printing/; "The UPS Store Makes 3D Printing Accessible to Start-Ups and Small Business Owners." 2013. *The Wall Street Journal,* July 31. Retrieved September 20, 2013, from http://online.wsj.com/article/PR-CO-20130731-911842.html#

6

ASSESSING PERFORMANCE FOR SUCCESS AND IMPROVEMENT

Opening Story: Lessons from the Steam Engine

October 25: After the Task-Force Meeting

"Paul, before you leave, could I talk with you for a moment?" David asked.

"Sure David, what's on your mind?" Paul responded.

"Your approach of 'taking out the trash' was quite novel. You dispelled false ideas and put the tough issues on the table. We really do speak different languages and we view IT's role differently. Acknowledging those facts set the stage for us to *confront the brutal facts*[1] about IT enablement of order fulfillment and customer service. Thanks for that," David said. Hesitating, David continued, "Anyway, as long as we are speaking bluntly, I've been losing sleep over Doug's first visit to Diane. When Diane called you to check on our delivery performance, you told her we were hitting on all cylinders. I've dug into the last two years' metrics. The metrics say your assessment was dead-on accurate. Our delivery is best in class. Monster's and Doug's reaction to the missed delivery window really rattled our world. We went from heroes to zeros in a single morning."

"That was painful for all of us," interjected Paul.

"That's my point. How could we be so naïve? We had no idea how ticked Monster was. We take pride in our metrics-driven decision making—and we were beating all of our targets."

"Yes, David, but we weren't hitting the customer's targets. I know where you are going with this. Measurement systems, like IT systems, are tools—they only do what we tell them to do. In this case, even though we've been tracking to industry standards, we simply haven't been measuring the right things. I guess that's now our job—to figure out what the right things are."

November 1: The Conference Room

"Good morning, everyone," David greeted. "Today, our task is to begin to discuss the role of measurement in supporting world-class delivery performance. You've seen the agenda. We want to start off conducting a blameless autopsy.[2] The issue at hand is simple: We were caught totally by surprise by Monster's harsh criticism of our delivery performance. Our culture is to measure everything, and we thought we were hitting the right targets. So, how is it that our measurement system failed in such a painful—and potentially costly—way? As we conduct the blameless autopsy, there are only two rules: (1) no holding back and (2) no finger-pointing."

"I'll be the scribe, if I can start," Lisé said as she stepped to the whiteboard (see Table 6-1). "In the spirit of brutal honesty—and taking out the trash—let me plead mea culpa. Over in finance, we put a lot of pressure on you guys to hit short-term, financial metrics."

"You sure do," agreed David. "You're always asking, 'What's the P&L impact?' Sometimes, we just don't know. A lot of what we do simply doesn't translate to P&L impact—at least not in the time frame finance seems to want. I hope that doesn't sound like finger-pointing."

"You're OK, David," Lisé replied. "Let's be honest, we have distinct world views. In finance, we answer to shareholders. But, our focus on the bottom line probably does handcuff you."

"It does! Of course, it doesn't help that we are managed as a cost center," David agreed. "It's tough to innovate and build new capabilities when we are always driving to reduce costs to keep top management happy. New capabilities cost money up front. The payback takes time."

"That's especially true if we don't get it right the first time—and we seldom do. It takes experimentation and a lot of changed behavior across functions. That takes time," Paul added.

"Lisé, you mentioned world views. So far, we've voiced three different world views. Let me add a fourth," Trina noted. "As a customer-facing profit center, marketing sees the world uniquely from each of you. Order fulfillment is a cross-functional process. Yet, if we all make decisions based on our local measures, we are going to create tradeoffs and maybe even conflict. The resulting chaos can cause us to lose sight of the customer and drop the ball."

"Fantastic," David exclaimed, "Your points are all on target. From a blameless autopsy perspective, they tell us that how we measure is 'killing us'—forgive the pun. When Diane asked me to head up this task force, I was also convinced that our problem was the 'how' of measurement. Until Doug dropped the hammer on us, I thought I understood why we measured—to document our performance. Right now, I'm not so sure about the why. Despite hitting our targets, we were failing to keep our most important customer happy. We really don't understand the ins and outs of our processes. Shouldn't measurement prevent this failing?"

"That's an astute observation," Paul chimed in. "After reading the agenda, I did a little out-of-the-box scanning. Do you know the story of the steam engine?" Blank stares prompted Paul to proceed, "The steam engine is the product of precise measurement. Only after finding a new way to measure the energy output of engines could inventors show that their ideas delivered better performance—you know, more power, less coal consumption. Without the micrometer, dubbed the 'Lord Chancellor'—which could gauge tiny improvements—the feedback needed to build better engines would have never emerged."[3]

"Let me jump even further afield. Did you know that Bob Beamon's long jump record of 29 feet, 2 ½ inches set in 1968 stood for almost 23 years? Mike Powell beat the mark by two inches in 1991. Nobody has come close since. Other track-and-field records come and go, but bettering jumping technique appears to be as hard as mastering order fulfillment. However, measurement is changing the science of jumping. BMW, as part of its London Games sponsorship, designed a camera system to give jumpers the immediate feedback needed to improve technique. Before a jumper leaves the pit, he knows his horizontal and vertical velocities as well as his flight angle. While the memory of the jump is still fresh, he finds out how what he did affected his performance."[4]

"Paul, where do you find time to read about steam engines and long jumping? Your point, however, is prescient. If measurement is going to help us build an outstanding order fulfillment capability, it must help us better understand and improve our processes. We should take a closer look at each of the issues on the board to see how accurate each point really is. I'll put my team to work on this, but that will take some time. For now, maybe we should turn our focus to addressing the question, 'What should a world-class measurement system look like?' Lisé, would you put that on the board? So, what do you think?"

Causes of Measurement Failure	A World-Class Measurement System Should...
Short-term emphasis on financial measures	
Tough to document unique value creation	
Too cost focused—sacrifice capability building	
Different world views; that is, conflicting measures	
Lose sight of customer; internally focused	
Poor understanding; no learning	

Consider as you read:

1. Based on your experience, which of the causes of measurement failure hinders outstanding order fulfillment and customer service the most?
2. How can you apply Paul's analogous discussion of steam engines (innovation) and long jumping (learning) to improve the design of your measurement systems?
3. How would you respond to David's final question: What should a world-class measurement system look like?

Assessing Performance for Success and Improvement

"Not everything that counts can be counted, and not everything that can be counted counts."
—Albert Einstein

A few years ago, the article, "On the Folly Rewarding A, while Hoping for B," appeared in print. What an intriguing title! Have you witnessed this phenomenon? Perhaps you are old enough to remember the U.S. auto industry of the 1980s. Having dominated car manufacturing for much of a century, GM and Ford found themselves under siege. Toyota and Honda had landed in America with an aggressive plan to capture market share—sell high-quality cars at a price lower than their American rivals. To borrow from the old idiom, the Japanese carmakers quickly began "kicking ass and taking names." Although GM and Ford attributed their Japanese rivals' success to "low-cost labor," the real secret to their success was a budding reputation for building reliable cars.[5] Toyota had established a *strategic breakpoint* in the area of quality.[6] The quality gap was so large that American consumers could not ignore it. By the 1990s, foreign nameplates—led by the Japanese—had captured over half of the California market.

How did GM and Ford respond? They announced (repeatedly) a new dedication to quality. In 1982, Ford aired a new commercial with the tagline, "At Ford, Quality is job 1."[7] But, nobody really believed it—and rightly so. Back on the production lines, productivity measures were still dominated. You know the rest of the story. Cost won out over quality! For the next 20 years, the Big Three U.S. carmakers struggled to close the quality gap and heal a tarnished quality image. Even today, many car buyers perceive Big Three products as inferior. What should you take away from GM and Ford's miscues? Measurement speaks louder—and more persuasively—than rhetoric. Steven Kerr, the author of "On the Folly Rewarding A, while Hoping for B," introduced the conundrum as follows:

> Whether dealing with monkeys, rats, or human beings, it is hardly controversial to state that most organisms seek information concerning what activities are

> rewarded, and then seek to do (or at least pretend to do) those things, often to the virtual exclusion of activities not rewarded. ...
>
> Numerous examples exist of reward systems that are fouled up in that behaviors which are rewarded are those which the rewarder is trying to discourage, while the behavior he desires is not being rewarded at all.[8]

Despite living in an age of enlightened management, you have probably seen this phenomenon repeated over and over again—and not just at work. For example, as a society, we constantly fret over the national budget deficit, but we incent our congressional representatives to pursue pork-barrel spending. We deplore our education systems' failures and our students' low scores on standardized tests. But, we continue to reward mediocre teaching and, amazingly, student GPAs continue to go up. The never-changing reality is that rewarding A while hoping for B will kill any effort to change—at every organizational level.[9] Rewarding A while hoping for B also alienates companies from critical supply chain stakeholders—customers, suppliers, and employees.[10]

The implication is clear. If you want to develop winning order fulfillment and customer service capabilities, you have to get measurement right. Measurement's power goes beyond motivation. Measurement provides the road map for helping you to go from your "as-is" capabilities to your desired "to-be" capabilities. Measurement also invites participation and incents alignment among all the diverse decision makers responsible for building—and constantly improving—those capabilities.

The Nature and Power of Performance Measurement

Attitude matters! Great companies are measurement fanatics. For example, one Fortune 500 executive expressed his company's attitude toward the measurement of order fulfillment activities as follows:

> If it moves we measure it. We measure for how much it costs to move, what resources were used, did we move it to the right place without damaging it and how long it took. If it doesn't move, we measure how long it stays there and what resources are consumed while it sits. This framework is applied to the measurement of products, people, and equipment. Finally, we try to measure whether we did it as well as or better than anyone else could do it.[11]

Why do great companies stress measurement? They know that sound measurement precedes the attainment of strategic goals—and ultimately success. You know the common catchphrase, "Information is power." Although many sources of information exist, measurement provides some of your most important decision-making information. However, unlike most sources of information, measurement informs both ways. How something is measured provides understanding of how—and how well—that something is working. Equally important, what is measured communicates the importance of that something, whether it is a strategy or an activity.

The nature and power of measurement can be grasped by digging a little deeper into three common measurement mantras:

- If you can't measure it, you can't manage it.
- What gets measured, gets done.
- Measure twice; cut once.

Measurement Informs Understanding

The time-honored adage, "If you can't measure it, you can't manage it" moves from trite to impactful when it is elaborated as follows: "If you can't measure it, *you don't understand it*, and therefore can't manage it!" Lord Kelvin said it this way, "When you can measure what you are speaking about, and express it in numbers, you know something about it; but when you cannot measure it, when you cannot express it in numbers, your knowledge is of a meager and unsatisfactory kind."[12]

The effective design and control of order fulfillment and customer service systems requires that you go beyond meager knowledge of their role and function. You must gain an intimate understanding of how—and how well—they work. In other words, you need quantifiable order fulfillment and service goals and the key performance indicators (KPIs) to inform them. Targeted and accurate measurement provides this understanding. Measurement reveals and details how the pieces of a system interact. It tells you how much progress you are making toward your fulfillment and service goals. When integrated with key partner systems, measurement informs you regarding (1) customer expectations and satisfaction levels as well as (2) supplier capabilities. And if a process (or relationship) is failing to deliver, a measurement-driven, problem-solving methodology will provide you insight into why things are not working. That is, measurement instigates the investigation needed to root out the causes of system failures. Measurement, thus, provides the feedback needed to develop and execute value-added strategies.

Measurement Motivates Behavior

An important question is often asked in boardrooms, on factory floors, and at the loading dock: "Does strategic intent really translate into results?" What is your take on this strategy-to-performance question? You may be surprised by the answer—or maybe not. Despite all the hours spent formulating and communicating strategy, logistics research suggests that the answer is often "No."[13] When does the strategy-to-performance connection break down? When measures are misaligned (see Figure 6-1). In the absence of supportive measures, strategy is not correlated with operational performance. Of course, this finding supports the introduction's assertion that it is folly to seek A while rewarding B. Tom Peters, McKinsey consultant and author of *In Search of Excellence,* expressed this empirical finding via the more memorable phrase, "What gets measured, gets done."

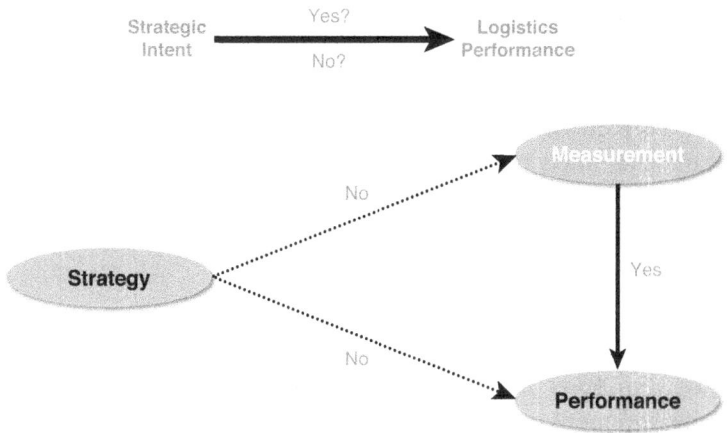

Figure 6-1   Connecting strategic intent to performance via measurement[14]

If you want to compete via order fulfillment and customer service, you need to remember that measurement shapes behavior. In fact, measurement is more influential than almost anything else you can do or invest in. You can hire the right people, provide them the right training, establish and articulate the right strategy, and still fail to perform if your measures promote counterproductive behaviors. People care about what is being measured. As a rule, people want to perform well. They want to be viewed favorably. They want to earn the merit raise. All of these outcomes hinge on their ability to perform according to what is being measured. When ego and pay are both tied to the same thing—that is, measurement—you can bet that people are paying attention. Because people want to know how they are being evaluated and they mold their expectations and behavior to the measures, you need to measure the right things.

Measurement Drives Execution

The first two mantras of measurement—that is, understanding and behavior—help you determine *what* to measure. The third mantra emphasizes *how* you measure. This idea is expressed by the motto, "Measure twice, cut once" found on the occasional coffee mug or T-shirt. If you have ever done a construction project (big or small) around the house, you've probably experienced the frustration this motto seeks to prevent. You measured and marked a board incorrectly, cut it to your measurement, and then put it in place—only to find that it did not fit. The experience taught you that it is imperative to measure the rights things *correctly*. An imprecise measure leads to a false cut, which costs you both time and materials.

If you measure the right things with right measures, good results naturally follow. Robert Kaplan, the architect of activity-based costing and balanced scorecards, has spent a lifetime trying to persuade decision makers that *how* they measure things matters. Elementally, he defines "correct" measurement as accurate, relevant, and timely.[15] If you have accurate, relevant, and timely information for every activity involved in order fulfillment and customer service, you will be able to make effective decisions and deliver outstanding value. By contrast, sloppy measures practically guarantee inefficient processes and dissatisfied customers.

To summarize, rigorous, thoughtful measurement always precedes great execution and world-class results. When your measurement system enables you to do the right things in the right way, your odds of winning go up dramatically. This reality is captured by the following measurement equation, which you should make visible throughout your organization:

$$\text{Understanding} + \text{Behavior} + \text{Execution} = \text{Winning Results}$$

An added benefit of getting the measurement equation right is that good measurement practice provides the feedback that is so essential to learning, innovation, and invention. William Rosen, author of *The Most Powerful Idea in the World: A Story of Steam, Industry, and Invention*, argues that without measurement-driven feedback, invention is "doomed to be rare and erratic."[16] The right feedback powers progress, making innovation "commonplace." In today's fast, clock-speed world,[17] outlearning rivals might be the only true source of sustainable advantage.[18] Good measurement practice can help you outpace your rivals in today's competitive race.

Measurement Practice—Understanding the Big Picture

Logistics is a measurement-rich environment. Traditionally, logistics measures have focused on five performance areas regarded as essential to delivering outstanding levels of profitable customer service. These five areas are asset management, cost, customer service, productivity, and quality (see Table 6-1).[19] Most companies use a variety of measures to help manage each value-added activity. Despite extensive use of diverse measures, many pundits (including Robert Kaplan) have long argued that traditional measures do not meet the needs of today's decision makers.[20] For example, Wickham Skinner cautioned that, "the very way managers define productivity improvement and the tools they use to achieve it push their goal further out of reach."[21]

Table 6-1   Traditional Measures Used to Manage Logistics Processes

Asset Management	Cost	Customer Service	Productivity	Quality
Inventory turns	Inventory carrying cost	Fill rate	Units shipped per employee	Damage frequency
Inventory obsolescence	Total landed cost	On-time delivery	Equipment downtime	Order entry accuracy
Return on assets	Outbound freight	Order cycle time	Order productivity	Picking/shipping accuracy
Inventory days supply	Warehousing labor costs	Complete orders	Warehouse labor productivity	Document/ invoicing accuracy
Economic value added	Administrative	Customer complaints	Transportation labor productivity	Number of customer returns

If you take a close look at the measures in Table 6-1, you will note that they tend to focus on the here and now; that is, they are short term and local in nature. As this is true across other functions within most companies, the quest for local excellence often leads to turf wars. Traditional measures simply do not provide a holistic view of value-added processes. The measures are also internally focused—and to a large extent reactive. As such, they do not provide visibility up and down the supply chain. Not only is it easy to lose sight of customer needs, but it is also almost impossible to anticipate their future aspirations. Further, a financial focus excessively weights efficiency, leading to a drive to cut costs rather than build capabilities. How would you use the measures in Table 6-1 to identify and document unique value-creation opportunities? Remember, it is almost impossible to "cut" your way to great order fulfillment. Finally, English physicist Lord Kelvin said, "If you cannot measure it, you cannot improve it."[22] Ask, "Do traditional measures enable or promote the fast-cycle organizational learning and innovation we need to win customer loyalty?"

Although not inherently wrong, much of traditional measurement practice impedes outstanding order fulfillment and customer service! Regrettably, when it comes to the design of measurement systems, the fact is that "insightful" measures have too often been sacrificed on the altar of "easy" measurement. For example, consider classic productivity measurement. What is your view on productivity growth? Most managers believe that productivity growth, like a higher ROA/ROI, is good—the higher, the better. The fact is, higher might be worse. Consider the ramifications of increasing order-picking productivity at the expense of poorer accuracy or higher damage incidence. If your ROI goes up because you skimp on needed investment, have you helped your firm position itself to win tomorrow's competitive battles? Not all tradeoffs are so easy to assess. With any of these measures, your real concern is why the measure (e.g., order-picking productivity) increased. If the process has improved because of better training, more clearly marked racks, or the use of radio-frequency technology, then you should document the improvements and make sure they are replicated across your organization as well as with key partners. If counterproductive behaviors are driving the so-called improved results, corrective action is needed. Productivity measurement can drive accelerated learning, but a narrow focus on efficiencies seldom does. More holistic, customer-focused measurement is needed.

Holistic Process and Supply Chain Measurement

A number of measurement-related processes facilitate true supply chain excellence for modern companies. These include total costing and the application of appropriate supply chain measures.

Total Costing

Total costing can reduce the propensity for counterproductive decision making described in the preceding discussion of productivity measurement. Total costs are simply the sum of all relevant costs for a given decision. Unfortunately, despite the rhetoric, few companies take a true total-costing approach to order fulfillment and customer service.[23] They can't! They do not know what all of the relevant costs are.[24] Figure 6-2 shares the findings from one research study, underscoring this reality. Ninety-four percent of logistics managers claim to use total costing, but in the next breath they acknowledge that they don't know what the cost of a back order or service failure is.[25] Consider how the problems associated with data availability and analysis are magnified as functional and company boundaries are crossed. Looking at Figure 6-3, how many decisions do you think are made using total supply chain costs? Does the lack of total costing influence service levels, customer satisfaction, and lifetime stream of profits?

To more fully understand the challenge of moving to more holistic costing practices, consider the notion of total cost of ownership (TCO), a type of total costing. TCO should be used in making purchasing decisions. It allows a true comparison of substitute materials or alternative suppliers. The equation for total cost of ownership is easy:

$$TCO = \text{Acquisition Cost} + NPV \, \Sigma \, (\text{Ownership Costs} + \text{Disposal/Scrap Costs})$$

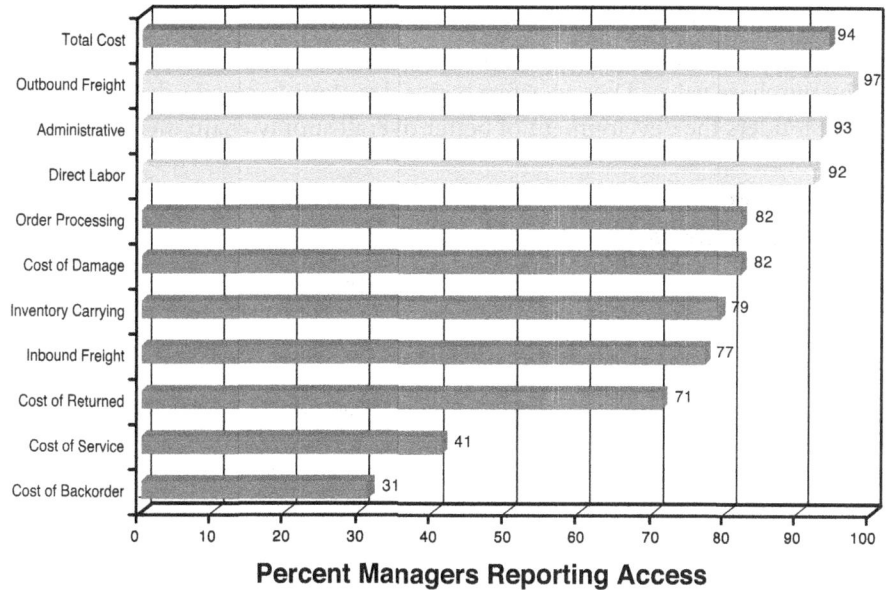

Figure 6-2 Data availability for total logistics costing

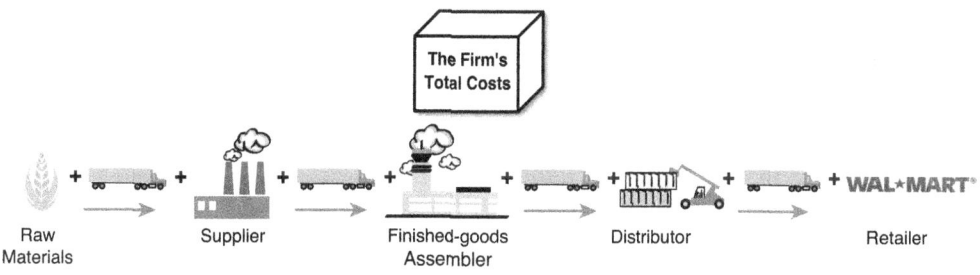

Total supply chain costs are the sum of all the costs incurred in planning, designing, sourcing, making, and delivering a product from raw materials to final customer.

Figure 6-3 Total supply chain costs[26]

The hard part is identifying all of the relevant costs. Figure 6-4 depicts the challenge. Some costs, like purchase price, transport costs, and duties, are easy to identify and tie to a specific purchase. Other costs, including material handling, yield issues, and field failures, are incurred postpurchase and are more difficult to quantify. For capital equipment, the total-costing challenge is even greater. Ownership costs must be estimated over a much longer period of time. The result: Managers make decisions based on the easy-to-identify information. They do not do the detective work to really understand the implications of their decisions. Given traditional reward systems, why should they? Figure 6-5 reveals that although many companies preach a TCO philosophy, what they measure is the purchase price. They sacrifice better decisions for easier decision making—a reality that hinders the development of better overall supply chain measures.

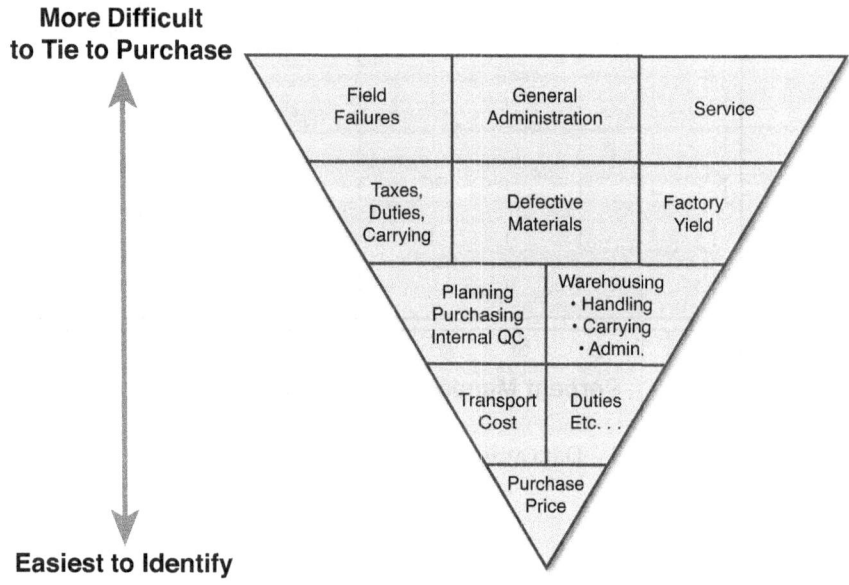

Figure 6-4 Total cost of ownership[27]

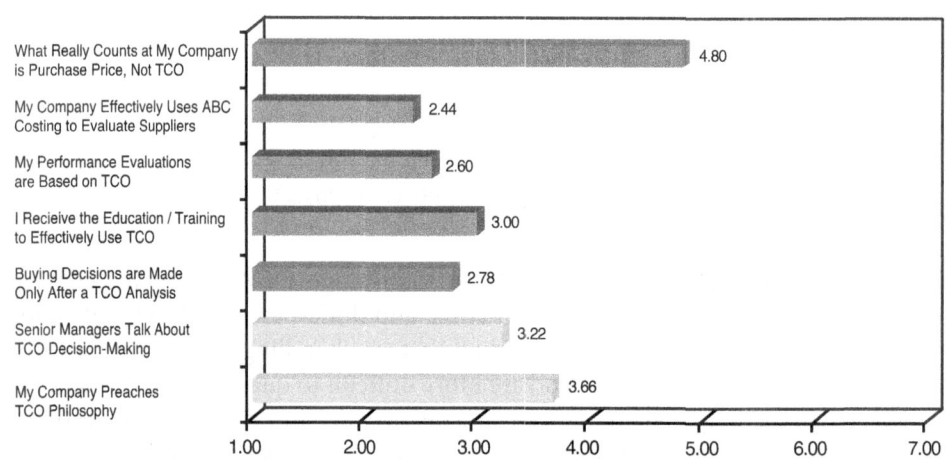

Figure 6-5 TCO building blocks

Supply Chain Measures

When asked to identify the key factors inhibiting better supply chain execution, one executive expressed a commonly held belief, saying, *"Metrics are critical! We don't know what the new ones should be, but we need them."*[28] The fact is that outstanding order fulfillment and customer service capabilities are supply chain capabilities. To provide fast, consistent cycles at the lowest costs, you have to rely on supply chain partners—namely upstream suppliers and logistics service providers—to perform. Yet, interorganizational (also known as supply chain–wide) metrics are almost nonexistent.

Over the past 20 years, beginning perhaps with the efficient consumer response initiative, many efforts have been undertaken to identify and define the supply chain measures of tomorrow.[29] Such measures, of necessity, are difficult to calculate and rely on information provided by supply chain partners. As a result, few breakthrough measures have been developed. Table 6-2 shows that these efforts have yielded three types of measures:

- **True supply chain measures**, that is, measures that do actually attempt to measure across the supply chain, but are so costly that they are only used as part of the efforts to benchmark supply chain performance for a specific industry. For example, the Efficient Consumer Response (ECR) initiative in the food industry documented over 120 days' supply of finished goods inventory was held by manufacturers, distributors, and retailers. Although some firms had reduced their own inventory levels, overall supply chain inventory remained high. Inventory had merely been shifted to other channel members.[30]

- **Theoretical supply chain measures**, that is, measures that express important concepts regarding supply chain performance and that would provide insight into interorganizational performance, but are not used to manage day-to-day operations.

- **Supply chain measures in name only**, that is, they are truly firm-specific measures, but they provide more insight into supply chain dynamics than traditional measures. Two of these measures have had noted and lasting influence on supply chain thought and practice—perfect order fulfillment and the cash-to-cash cycle time:

    - *Perfect order fulfillment*—Perfect orders are hassle free. You can describe this in one of two ways: everything was just right or nothing went wrong. Consider how hard this is to achieve. Gus Pagonis, the general who managed logistics and resupply for the first Gulf War, described logistics as follows: "If you go an hour without a screw-up, you've had a great day." The perfect order thus sets a very high performance bar. The Supply Chain Council defines a perfect order as follows: an order delivered to the right place, with the right product, at the right time, in the right condition, in the right package, in the right quantity, with the right documentation, to the right customer, with the correct invoice.[31]

Table 6-2 Supply Chain Performance Measures

Terminology	Definition
True Supply Chain Measures	
Inventory days supply	The total number of days of inventory required to support the supply chain—from raw materials to the final customer acquisition. This is expressed as calendar days of supply based on recent actual daily cost of sales.
Total supply chain cost	The sum of all the costs incurred in planning, designing, sourcing, making, and delivering a product broken down for each member of the supply chain.
Theoretical SC Measures	
Supply chain response time	The theoretical number of days required to recognize a major shift in market demand and increase production by 20 percent.
Source/make cycle time	The cumulative time to build a shippable product from scratch—if you start with no inventory on hand or on order. It consists of total sourcing lead time, release-to-start build, total build cycle time, and complete build-to-ship time.
Value-added productivity	Total company revenues generated less the value of externally sourced materials expressed as a ratio of total company headcount.
SC Measures in Name Only	
Perfect order fulfillment	A perfect order is an order that is delivered complete, on time, in perfect condition, and with accurate and complete documentation. Fulfillment is the percent of orders that are perfect (Perfect Orders/Total Orders).
Cash-to-cash cycle	The time required to convert a dollar spent to acquire raw materials into a dollar collected for finished product (Total Inventory Days of Supply + Days Sales Outstanding–Days Payables Outstanding).
Order fulfillment cycle time	The average actual lead times consistently achieved, in calendar days, from customer order to customer delivery. This includes order authorization to entry, entry to release, release to shippable, shippable to customer receipt, and receipt to customer acceptance.
Inventory dwell time	The ratio of days inventory sits idle to days inventory is being productively used or positioned.
On-shelf in-stock percentage	The percentage of time that a product is available on the shelf, rack, or wherever the customer expects to find and buy it. This measures the supply chain's ultimate ability to satisfy the end customer.
Customer inquiry response time	The average elapsed time between receipt of a customer call and connection with the appropriate company representative.
Customer inquiry resolution time	The average elapsed time required to completely resolve a customer inquiry.

Any kind of error busts a perfect order. Nine out of ten imperfect orders fail due to one of the issues listed in Table 6-3. Outstanding performers hit perfect order percentages of only about 90 percent. The Grocery Manufacturers Association provides performance statistics for a few components of a perfect order (see Table 6-4). Focusing just on four issues—on-time delivery, orders shipped complete, orders shipped damage free, and orders shipped with the correct documentation—the average perfect order percentage is 84 percent.[32] Pursuing the perfect order can help you avoid complacency and drive improvement as you find and remove the root causes of perfect order failures. Until recently, many considered the perfect order to be the pinnacle of logistics performance. However, some leading firms see a flaw in this logic: The perfect order is inward looking; it ignores customer perceptions of service levels.

Table 6-3 Perfect Order Busters

Order entry error	Missing information
Ordered item is unavailable	Late shipment
Incomplete paperwork	Inability to meet ship date
Picking error	Early arrival
Customer deduction	Inaccurate picking paperwork
Damaged shipment	Invoice error
Overcharge error	Credit hold
Error in payment processing	

Table 6-4 Calculating a Perfect Order

Perfect Order Elements	Average
Percent on-time delivery (OTD)	93 percent
OTD to DC within 30 minutes	
Percent orders shipped complete	97 percent
(order line item fill rate)	
Percent of orders shipped damage free	99 percent
Percent of orders sent with the correct documentation	94 percent
Invoice accuracy	
Total Perfect Order (.93 × .97 × .99 × .94 = .84)	84 percent

- **Cash-to-cash cycle time**—Cash is king. Recognizing this fact, companies are managing their cash flow more aggressively. One CFO explained the strategy as follows:

  > We have made working capital reduction a priority. ...We are able to operate some of our businesses with negative working capital. Rather than putting money in inventory or receivables we prefer to have our suppliers finance us by increasing our short-term liabilities, thus freeing up capital for other investments.

  How much money can a company free up? A recent *Wall Street Journal* article noted, "Procter & Gamble is planning to add weeks to the amount of time it takes to pay its suppliers, a shift that could free up as much as $2 billion."[33] The opportunity to "live off of someone else's dime" has made cash-to-cash cycle time a pivotal measure—one that shows the direct financial benefit of supply chain initiatives to senior management. This is critical as finance is the language of business. Freed-up working capital can be invested in new products, better processes, or more collaborative relationships.

  The formula to calculate a company's cash-to-cash cycle is as follows:

  Cash-to-Cash Cycle = Total Inventory Days of Supply

  + Days Receivables

  − Days Payables

  Let's use this equation to look at two companies' cash-to-cash performance over time. Dell and Walmart are often considered to be cash-to-cash champions. Their cash-to-cash calculations for 2004 and 2012 are shown in Table 6-5. A close look shows a company's cash-to-cash performance depends on two key issues: (1) highly efficient processes that minimize inventory and speed collection of payables and (2) leverage with supply chain partners that enables quick collection of receivables and delayed payment to suppliers. From a co-value-creation perspective, building efficient processes is the preferred approach to reducing cash-to-cash cycles. How do Dell and Walmart stack up? Both companies achieve short or negative cash-to-cash cycles, granting them financial flexibility. Both companies rely on extremely tight logistical operations to minimize inventory. Dell leans more on leverage to extend its payables. Walmart collects receivables more quickly.

Table 6-5 Cash-to-Cash Cycle Calculations for Dell and Walmart*

	Sales	Inventory	Receivables	Payables	Inventory Days	Days Receivables	Days Payables	C2C Cycle
2004								
Dell	$39,667	$327	$3,635	$9,935	3.01	33.45	91.42	−54.96
Walmart	$258,681	$26,612	$1,254	$31,051	37.55	1.77	43.81	−4.49
2012								
Dell	$62,071	$1,404	$6,476	$11,656	8.26	38.08	68.54	−22.20
Walmart	$446,114	$43,803	$6,768	$38,080	35.84	5.54	31.16	10.22

* Dollar values expressed in millions.

Finally, although most companies use aggressive cash management to free up working capital, a few have identified an opportunity to improve supply chain relationships and overall supply chain performance. One executive explained his company's novel approach to cash management, "Most large companies strive to reduce their cash-to-cash cycle, but not every company has the same cost of capital. Total supply chain costs can be reduced if the company with the lowest cost of capital accepts longer cash-to-cash cycles. We try to look at this as we manage relationships."

Customer-Centric Measurement

A new metric is gaining attention. Although its acronym—SAMBC—isn't nearly as catchy as "the perfect order," the concepts it promotes are catching on as a better way to evaluate order fulfillment and logistics service performance. Deirdre White, associate director of customer service at P&G, defines Service As Measured By the Customer as, "The percentage of measured customers at which we are at or better than expected service targets, where the targets are established by and with each customer." Ms. White explains why SAMBC is a more relevant and important metric, "Our view on service had been very internally focused. We lost a lot of opportunity to create value for our customers and ourselves."[34]

Regardless of how well your company thinks it is performing based on the internal metrics, what really matters is how the customer perceives your performance. That is, your customer gives you the grade that matters, deciding whether you pass or fail. Using the example of a supplier that splits a 100-case shipment (95 shipped/5 short) because of space constraints, Ms. White demonstrates how easy it is to see past each other in the order fulfillment process. The supplier ships the remaining cases on the next truck and feels it has fulfilled the order. As you might guess, the customer disagrees.[35] Unhappy, the customer is likely to look for a new, more perceptive and responsive supplier.

To meet customers' real needs, you must understand them from the customer's vantage point (see the discussion in Chapter 1, "Meeting Customers' Real Needs: The Nature of Service System Design," regarding customer service, satisfaction, and success). Traditional customer service measurement—which relies on internally generated statistics on such things as fill rates, on-time delivery, and response time to inquiries—cannot provide this insight. You have to ask customers what services they truly value. Then you have to find out how they measure them. You may discover that the "high" service levels you have been providing are not highly valued by customers. P&G's White noted customer conversations can reveal nasty surprises. She explained that in one instance, it dawned on members of the fulfillment team that P&G had "missed expectations for several years." After digging into the problems and better defining who controlled each part of the fulfillment process, P&G's performance improved by 50 percent within six months.[36]

Establishing customer-centric measurement is more about process than building a list of metrics. The process requires that you not only meet with individual customers to agree on the metrics to be used, but also align your measures to the customers' measures—at least for your most important customers. Even though many metrics are industry standards, each customer is likely to have tweaked them to fit his or her own competitive priorities. At UPS, customer feedback led managers to change measures that were expressed in terms of average performance levels (e.g., percent of shipments delivered damage free) to absolute measures (e.g., damaged shipments delivered to a specific customer). A senior manager explained, "To say that we deliver 99.5 percent of our packages damage free might give us a false sense of well-being. To say that 5,000 customers received damaged packages from us on a particular day puts an entirely different perspective on our performance." Table 6-6 summarizes the essence of customer-centric measurement.

Table 6-6   Customer-Centric Measurement

Traditional Customer Measurement	Customer-Centric Measurement
Internal service measures over satisfaction measures	External assessment that reveals what customers really think is important and how they measure your performance
Measures that are expressed as averages	Measures that are expressed in both average and absolute terms—on a customer-by-customer basis
Measures based on industry standards and that treat all customers the same	Measures that recognize unique needs of individual customers and are aligned to those needs

Balanced Scorecards

The early nineties were a time of rapid change. The competitive environment was more dynamic and turbulent than ever before and new technologies were enabling companies to experiment with new business models. However, managers felt they lacked the measurement capability to drive needed change and implement new strategies. In fact, many managers felt that measurement was holding them back—keeping them from making decisions that would create the value customers were demanding. Within this setting, Robert Kaplan and David Norton introduced the balanced scorecard concept. Their goal was to provide a measurement framework that would overcome the backward-looking and short-term nature—which were viewed as critical deficiencies—of widely used financial metrics.[37]

Why was a new, balanced approach needed? In an environment of constant change, managers needed a tool that could both inform and motivate the right change. Remember, measurement is powerful precisely because it informs decision making and motivates needed behaviors. Larry D. Brady, executive vice president at FMC, commented on the need for better measurement, saying, "If you are going to ask a division or the corporation to change its strategy, you had better change the system of measurement."[38]

The balanced scorecard supports change and better decision making by incorporating four distinct dimensions into a holistic management dashboard. Specifically, a balanced scorecard invites you to ask four questions about your company's mission and strategy:

- "To achieve our vision, how should we appear to our customers?"
- "To succeed financially, how should we appear to our shareholders?"
- "To satisfy our shareholders and customers, what business processes must we excel at?"
- "To achieve our vision, how will we sustain our ability to change and improve?"[39]

By focusing on customer expectations, operational excellence, future capability development, and financial metrics, the scorecard brings balanced insight to decision making (see Figure 6-6).

Taking a balanced approach yields several advantages vis-à-vis a more financially oriented measurement methodology. First, scorecards translate vision into actionable behavior. The scorecarding process provides a top-down reflection of your company's goals, deriving measures directly from strategy-driven goals. Second, the resulting scorecard communicates top-management's vision throughout the organization. Research has shown that when a scorecard is not in use, up to 95 percent of a typical workforce cannot articulate the company's strategy.[41] Members of your team cannot align to and coalesce around a strategy they do not understand. Third, a well-designed scorecard not only integrates internal and external perspectives, but it is also forward looking. Your scorecard invites managers to ask, "What capabilities do we need to build to better meet customer needs and improve shareholders' returns?" Naturally, the follow-up question is, "How do we invest in and organize for this capability development?" Fourth, because a complete scorecard should consist of only about 20–25 metrics, it helps you focus on what is really strategically important.[42] The scorecard imposes discipline, helping decision makers avoid distractions and "shiny objects."

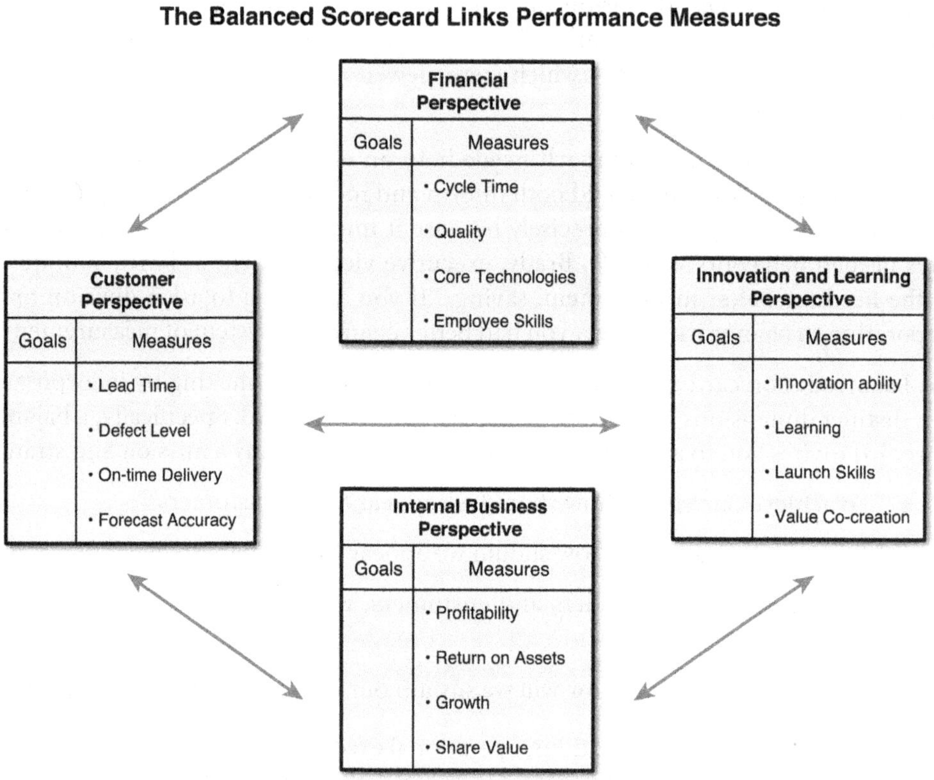

The Balanced Scorecard Links Performance Measures

Financial Perspective

Goals	Measures
	• Cycle Time
	• Quality
	• Core Technologies
	• Employee Skills

Customer Perspective

Goals	Measures
	• Lead Time
	• Defect Level
	• On-time Delivery
	• Forecast Accuracy

Innovation and Learning Perspective

Goals	Measures
	• Innovation ability
	• Learning
	• Launch Skills
	• Value Co-creation

Internal Business Perspective

Goals	Measures
	• Profitability
	• Return on Assets
	• Growth
	• Share Value

Figure 6-6   The balanced scorecard[40]

Now, the pivotal question, how do you construct an effective scorecard? The process begins with a firm and determined recognition that a new approach to measurement is truly needed. If your team doesn't buy in to a deep underlying need for a scorecard, they will not invest the time and effort to do it right. Because scorecards are so powerful at communicating strategy and guiding behavior, a bad scorecard is worse than no scorecard. Figure 6-7 provides the basic blueprint for scorecard construction. Once your vision of the future is defined and agreed upon, you must translate vision into strategy. Ask the question, "If my vision succeeds, how will I differ to my shareholders, my customers?"[43] This question will help you determine the nature of your "to-be" value-added and innovation capabilities. With this high-level foundation in place, the process becomes much more granular. The following four steps will help you translate strategy into an actionable scorecard:

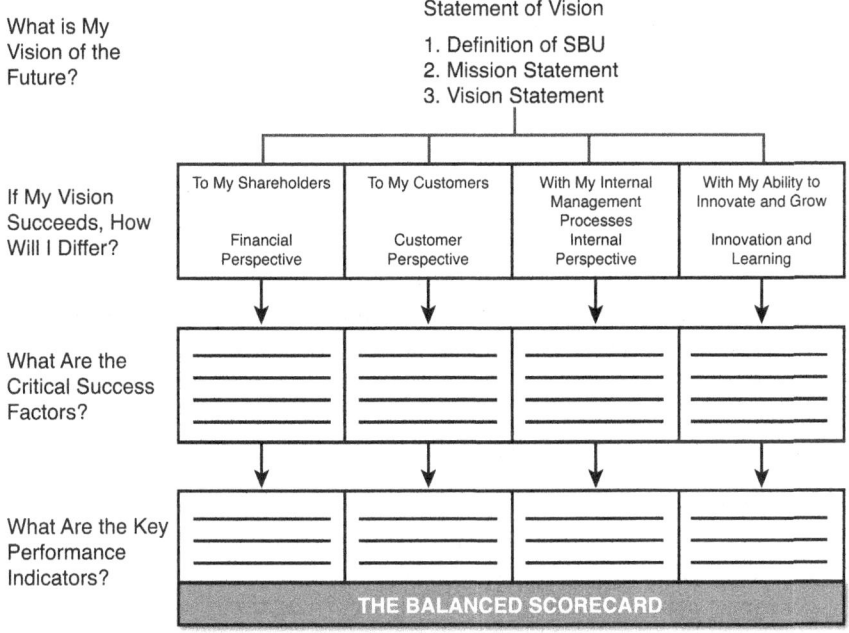

Figure 6-7 A blueprint for constructing a balanced scorecard[44]

1. **Establish goals**—Goals matter! They set the direction for your company and they define what you need to measure; that is, they define the understanding, motivation, and execution you need your metrics to promote. To make sure you get your goals right, you should consult the following sources of information:

    - **Environmental scan (also known as SWOT analysis)**—Most companies conduct a periodic scan of the competitive environment to identify potential competitor moves, changes to industry structure, emerging technologies, and future government regulations. Each of these issues could potentially change the rules of the competitive game, requiring you to build new capabilities in order to survive and thrive.[45]

    - **Best-in-class benchmarks**—If you want to offer customers truly competitive service levels, you need to benchmark best practices for relevant capabilities and processes. Benchmarking is the formal process of comparing the attributes of one organization with those of another. Proactive benchmarking helps you assess and improve your company's competitive abilities. Benchmarking involves the following:

a. Identifying the attribute (e.g., capability, process, or routine) you want to improve and then finding a company that achieves best-in-class performance in that area

b. Documenting the benchmark company's process at strategic and operational levels to identify specific opportunities for improvement

c. Developing a detailed, step-by-step strategy to bring the best practice in-house

Your quest is to find the very best ways to do the things (e.g., fast-cycle delivery, vendor managed inventory, integrated business planning) that matter to your customers. To do this, you have to banish the not-invented-here mind-set that afflicts many firms. Your motto should be, "Good practice is good practice, wherever it is found." Deliberate benchmarking not only improves specific capabilities, but also stimulates constant learning.

- **Customer feedback**—If you look at the mission statements of successful companies, you will consistently find five essential words, "meet or exceed customer expectations." To do this well, you need to talk with your customers. To get the real skinny on how you can better meet their needs, you need to earn your customers' trust. This means they trust you will be able to live up to your promises of better service and they want to work with you over the long haul.[46] Otherwise, they are unlikely to waste their time giving you the feedback you need to really distinguish your performance.

2. **Identify metrics**—With goals established, you are ready to derive the metrics that make up the heart of your scorecard. In addition to evaluating current measures and conducting internal brainstorming sessions to ideate new measures, you should look to the following for inspiration:

- **Industry standards**—In most industries, a set of standard metrics exists for key value-added activities. For logistics, standard metrics are likely to look like those listed in Table 6-1. Professional associations, industry consortiums, consultants, and textbooks are good places to look to make sure you do not miss the obvious.

- **Best-in-class benchmarks**—Just as some companies excel at value creation, others are known for their use of creative and effective measurement practice. Check out what they are doing and see how their metrics fit your needs.

- **Customer feedback**—When you ask customers how you can better meet their needs, follow up by asking how they evaluate your service. Request a copy of their supplier scorecard and go over it with them to find out exactly how they define and calculate their measures. This understanding will help you align your measures with theirs.

Because your competitive context, core competencies, and customer relationships are distinct, you will want to tailor some of your metrics. Tailored metrics recognize the fact that a one-size-fits-all approach to measurement cannot realistically cultivate high-caliber, value co-creation relationships.

Tailored metrics derive their value from their ability to communicate specific expectations, align partner efforts, track specific process capabilities and performance, and promote the collaborative behavior needed to pursue special projects and inculcate one-of-a-kind relationships. Table 6-7 shares a checklist of the attributes of an effective tailored measure. For each "No" that you check, you need to ask if this deficiency will undermine the metric's effectiveness, leading to poor understanding, counterproductive behavior, and unintended consequences.

Table 6-7 A Checklist for the Development of Tailored Metrics[47]

This Tailored Measure Is...

Yes	No	
☐	☐	Aligned with organizational goals
☐	☐	Aligned with project goals
☐	☐	Customer oriented
☐	☐	Meaningful to workers, managers, and customers
☐	☐	Consistent across appropriate functions or departments
☐	☐	Designed to promote cooperative behavior both horizontally and vertically
☐	☐	Communicated to all relevant individuals
☐	☐	Simple, straightforward, and understandable
☐	☐	Easy to collect the needed data
☐	☐	Easy to calculate
☐	☐	Available on a timely basis—in real time when possible
☐	☐	Strategic and tactical
☐	☐	Quantifiable
☐	☐	Designed to drive appropriate behavior
☐	☐	Designed to drive learning and continuous improvement
☐	☐	Designed to provide information that is actually used in decision making

Adapted from Fawcett, Ellram, and Ogden (2007)

3. **Integrate with information system**—To be effective, your scorecard needs to be populated with accurate, up-to-date information. During the scorecard development process, you need to make sure that your systems are able to collect and analyze all of the information needed to populate the scorecard. You also need to run a pilot test to ensure that all the links are correct and the metrics are valid. GIGO can sully the credibility of a newly launched scorecard.

4. **Implement**—When the scorecard has been fully validated, you need to make sure that it is rolled out with proper training. The key here is to make sure people understand why the scorecard was developed, what individual metrics mean and how they are calculated, and how to use it for holistic decision making. The odds are that many of the decision makers who will be using the scorecard should have participated in its development. Although such inclusion may increase development time, it mitigates implementation headaches. People support what they help create.

To summarize, the scorecarding process unleashes the power of measurement. A well-designed scorecard helps you develop the strategic focus, the value-added capabilities, and the supply chain relationships you need to win tomorrow's battles for the customer's loyalty. It also accelerates learning. Finally, when customized scorecards are co-created with your most important customers, you achieve better alignment, build deeper and stronger trust, and set the stage for future collaboration.

Measurement Practice—Delving into the Details

Driving operational excellence means getting into the trenches to learn how—and how well—the specific activities and processes that constitute order fulfillment and customer service are really working. Figure 6-8, Panel A, shows a format used by many companies to disaggregate core elements of logistics service into metrics. Adopting a similar process will help you measure the little things that ultimately determine whether or not a process performs as promised. Panel B provides standard definitions for key fulfillment and service metrics.

Paying close attention to these measures underlies your ability to control daily operations and promote continuous improvement. The feedback from these measures can help you answer the following question: What should we be willing to pay to improve performance? Let's take a look at two fulfillment goals—providing outstanding product availability and shortening order fulfillment cycles—to illustrate the type of analysis you need to conduct to evaluate performance and profitability tradeoffs.

Translating Service Goals to Metrics		
Element	Definition	Typical Metric Measured from Customer Perspective
Product Availability	Usually defined as percent of times product is available to fill first request orders	• Order fill rate • Case fill rate
Order Cycle Time	Activities and time that elapses between when an order is placed and when the shipment is received.	• Order cycle time in days • % orders received within x days • On-time delivery
Logistics Operations Responsiveness	• Ability of the supply chain to meet special customer requests • Ability of the supply chain to adapt to sudden changes in volume	• Response to special request (hours, days) • Time required (days) to respond to x% increase in unanticipated demand
Logistics System Information	Ability of an information system to supply timely and accurate information	• Response time (hours, days) to requests for information • % compliance to EDI
Postsale Product Support	Ability to provide customer support after product delivery, including technical information, spare (replacement) parts, or product return	• Response time to service request • Cost to return product • Spare part availability

Source: Adapted from James F. Robeson and Robert G. House

Availability Measures	
Fill rates	Measures the magnitude or impact of stockout over time.
Stockout frequency	Refers to the probability that a stockout occurs in a firm with no available inventory to meet customer orders.
Orders shipped complete	Requires shipping everything that a customer orders to count as a complete shipment.
Operational Measures	
Speed of the performance cycle	The elapsed time from when a customer established a need to order until the product is delivered.
Consistency of the order cycle	Measured by the number of times that actual cycles meet the time planned for completion.
Malfunction recovery	The ability to quickly implement contingency plans when a failure occurs in the supply chain.
Service Reliability Measures	
Damage-free shipments	Measures how many shipments arrive without damaged products.
Error-free invoices	Measures what percentage of invoices contain no errors.
Shipment/order match	Order measures how many shipments contain the exact amount of product ordered.
Shipped to correct location	Measures how many shipments are made to the customer's selected location.

Source: Definitions taken from Bowersox, Closs, & Cooper, (2012)

Figure 6-8   Logistics order fulfillment and service metrics[48, 49]

Product Availability

Product availability is typically measured using a fill-rate metric; that is, item fill rate, line fill rate, or order fill rate. Figure 6-9 illustrates how these metrics are calculated for an order that consists of multiple lines. Please note that the customer has ordered different quantities for each line (i.e., a specific SKU). As a result, the choice of metric—item versus line fill rate—provides a very different perspective on the level of service provided. For Scenario A, the line fill rate is a respectable 90 percent. However, the item fill rate is 45 percent—a truly dismal service level. The math for Scenario B shows an item fill rate of 69 percent and a line fill rate of 40 percent. Although the item fill rate is better than the line fill rate, both represent unacceptable performance. For both scenarios (A & B), the order fill rate is 0 percent, so is the perfect order rate! The point is simple: You need to understand the nature of the metric, how it is calculated, and what the results are truly telling you. This fact is true for all metrics, not just fill-rate measures. You can't afford to let the inappropriate use of metrics (either wrong metrics or wrong use of metrics) lull you into a false sense of high performance.

Order Consists of Multiple Lines			
Line	Items Ordered	Scenario A Items Filled	Scenario B Items Filled
1	10	10	0
2	20	20	0
3	30	30	30
4	10	10	0
5	20	20	0
6	30	30	30
7	10	10	0
8	20	20	0
9	30	30	30
10	220	0	220
Total	400	180	310

Item Fill Rate
Scenario A: 180 ÷ 400 = 45%
Scenario B: 310 ÷ 400 = 69%

Line Fill Rate
Scenario A: 9 ÷ 10 = 90%
Scenario B: 4 ÷ 10 = 40%

Figure 6-9 What do different fill-rate metrics mean?[50]

Adapted from Coyle, Langley, Gibson, and Novack (2011)

Now, let's address the question of how much you should be willing to pay to improve availability performance. Consider the ensuing scenario, which might describe your last year's order profile:

- The average order is for 1,000 units.
- You delivered 10,000 orders last year.
- Your pretax profit per unit was $50.
- Your pretax profit per order was $50,000.
- Your percent of incomplete orders back-ordered is 70 percent (30 percent are canceled).
- Your back-order costs per order were $200 (e.g., administrative, handling, shipping).
- Invoice deduction for short orders is $500.

If you delivered 85 percent complete orders last year, how much cash flow did you lose because of poor product availability (i.e., your inability to ship complete orders)? The *relevant costs* to include in the calculation are as follows:

- Cost of canceled orders (i.e., 30 percent of incomplete orders × pretax profit per order)
- Back-order costs (i.e., 70% of incomplete orders × back-order cost)
- Invoice deduction cost (i.e., 70% of incomplete orders × invoice deduction)

Based on the 85 percent complete order rate, your total number of incomplete orders was 1,500 (i.e., 15% × 10,000 orders). The calculations follow:

$$\text{Cash Flow Lost} = \$23{,}235{,}000 \quad \text{or} \quad \begin{array}{ll} .3 \times 1{,}500 \times \$50{,}000 & \text{(canceled orders)} \\ + .7 \times 1{,}500 \times \$200 & \text{(back-order cost)} \\ + .7 \times 1{,}500 \times \$500 & \text{(deductions)} \end{array}$$

Suppose that during your visits to customers, they complained that you were 10 percent under the industry standard for shipping complete orders. They implored you to do something about this. How much money should you be willing to invest in added inventory, better information systems, or closer relationships (e.g., VRM) that would enable you to hit the 95 percent target?

Based on a 95 percent complete order rate, your total number of incomplete orders would be 500 (i.e., 5% × 10,000 orders). Your new lost cash flow is calculated as follows:

$$\begin{aligned}
\text{Cash Flow Lost} = \quad \$7{,}745{,}000 \quad \text{or} \quad &.3 \times 500 \times \$50{,}000 && \text{(canceled orders)} \\
+ &.7 \times 500 \times \$200 && \text{(back-order cost)} \\
+ &.7 \times 500 \times \$500 && \text{(deductions)}
\end{aligned}$$

The low end of what you should be willing to pay to improve fill-rate performance is the difference between $23,235,000 and $7,745,000 or $15,490,000. This $15.5 million becomes the number against which you weigh the costs and benefits of different improvement initiatives. Why is this the low end? The odds are pretty good that if you improve your ability to deliver full shipments, two things will happen. First, your customers will cancel fewer orders, reducing your canceled order costs. Second, your better performance will attract some additional sales, increasing your top-line revenues.

Order Cycle Time

The ability to deliver orders quickly can buy you flexibility in how you manage your overall value-creation system. That is, if you have a fast-cycle logistics delivery capability (Amazon's rationale for building a vast network of fulfillment centers), you could potentially smooth your production flow, allowing you to lengthen production runs, eliminate wasteful setups, and focus on innovating more-effective processes. Of course, you might simply seek to capture market share by offering faster-than-industry-average delivery times, which would reduce your customers' needs to carry inventory.

The question is: How much should you be willing to pay to improve performance? You know that the basic order cycle consists of four major components: order placement, order processing, order preparation, and order shipment. Each component—as well as the overall order cycle—can usefully be described statistically using the mean and range. That is, your customers want to know when to expect the order to arrive (the mean) and what the worst-case scenario is (the upper end of the range). Figure 6-10 depicts these relationships.

Now, imagine you have met with your customers to learn more about their expectations as well as their sensitivity to faster order cycle times. They tell you that shorter cycles are important. Indeed, they would like you to reduce your current order lead time by 50 percent. They would also like you to reduce the variability of the order cycle; that is, they want you to promise a more consistent, reliable lead time. As you talk, it is clear that customers feel the current order cycle is driving up two inventory-related costs: (1) demand inventory to cover the order cycle lead time and (2) safety stock to compensate for your highly variable order cycle. Subsequent to meeting with your key customers, you mapped out your order cycle to find out how it was working. You also initiated a series of brainstorming activities to see if a 50 percent reduction in cycle time is even possible—let alone financially viable. The facts you discovered are as follows:

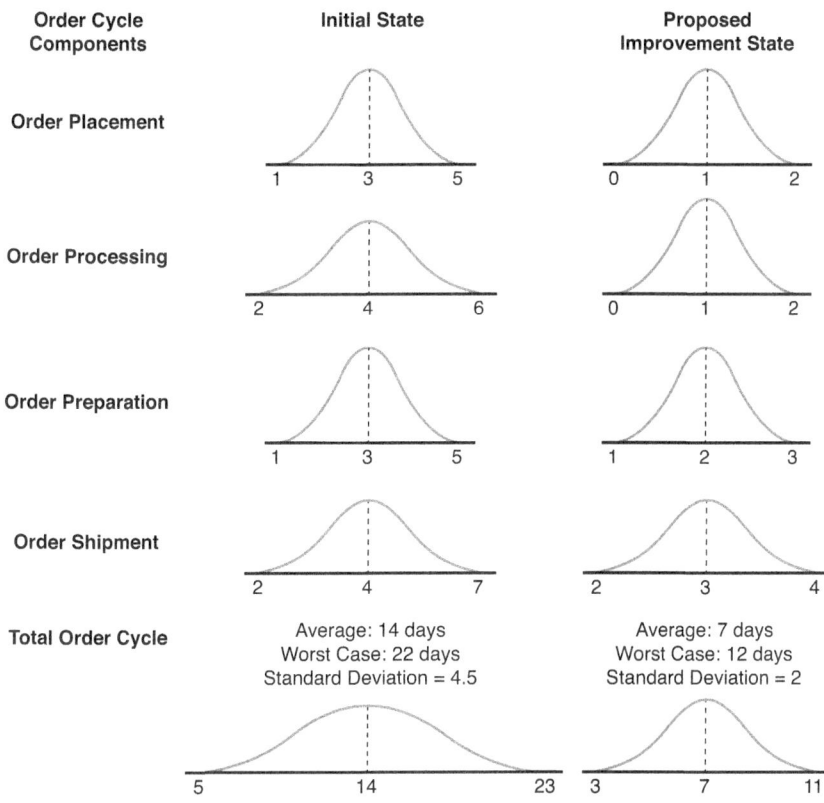

Figure 6-10   Understanding the effects of order length and variability[51]

	Current Customer	Capability Request
Average order cycle time	14 days	7 days
Standard deviation of cycle time	4.5 days	2 days
Service level	95 percent	

You also know the following for your most important customer:

- Sales price: $250
- Historical demand per day: 5,000 units
- Inventory carrying cost: 20 percent
- The z-score for a 95 percent service level: 1.96

Based on the nature of the system and its costs, you are confident the greatest opportunity for cost avoidance is found in the demand inventory for your 14-day lead time. By cutting your lead time in half to 7 days, your customer could reduce its inventory levels by 7 (i.e., 14−7=7) days' worth of sales. Thus, your calculations are as follows:

Inventory Cost Savings = 7 days × 5,000 units/day × $250 per unit × 20% = **$1,750,000**

Turning to the safety stock calculations, you decide you want to make sure everyone sees the actual inventory impact of the proposed change. So, you calculate the absolute unit numbers for current versus proposed safety stock levels. Figure 6-10 modeled your order fulfillment system statistically, creating a standard normal probability distribution for lead-time performance. Variability is measured by the standard deviation of cycle time. Your confidence that the system will perform is captured by the z-score. From a statistical perspective, the z-score is the distance from the mean as measured by the number of standard deviations. A z-score of 1.96 signifies that you are 1.96 standard deviations above the mean. That is, your fulfillment system will deliver to this time 95 out of 100 times. Your calculations follow below:

Current Safety Stock = 5,000 units/day × 1.96 × 4.5 days = 44,100 units

Proposed Safety Stock = 5,000 units/day × 1.96 × 2 days = 19,600 units

Thus, if you can reimagine your order fulfillment system so that it delivers in 7 days on average with a standard deviation of 2 days, your customers will need to carry 24,500 fewer units of safety stock inventory (i.e., 44,100−19,600). The cost savings are calculated below:

Safety Stock Cost Savings = 24,500 units × $250 per unit × 20% = **$1,225,000**

The total inventory cost savings for your customer is the sum of the demand and safety stock savings:

Total Inventory Cost Savings = $1,750,000 + $1,225,000 = **$2,975,000**

If your customer is willing to share cost savings on a 50/50 basis for the first year, you could invest up to $1,487,500 to improve your system. P&G has provided up-front funding for smaller suppliers to make such investments because managers at P&G know that improved supplier performance will enable P&G to recoup the investment many times over (and year after year). Of course, shorter cycle times might also reduce the amount of inventory you have to hold in order to meet customer service expectations. This would free up additional capital to invest in cycle-time reduction. Finally, shorter cycle times could improve your value proposition, helping you capture market share and driving top-line revenue gains. But, you won't know until you find out what customers want, model your system, and crunch the numbers.

Your takeaway: The devil is in the details. So are your opportunities to drive consistent improvements in your order fulfillment and customer service processes. Well used, metrics can drive the constant learning and improvement that is needed to win tomorrow's competitive game.

Conclusion

Measurement matters! Consider the following examples:

- A logistics professional in the midst of a commuting nightmare came home each week to a discouraged and distraught sweetheart. After returning home one Friday night and listening to all sides of the story, he grasped that without more help around the house, the mess and chaos had become overwhelming. Knowing that good practice is good practice, he developed a simple scorecard, defining goals and establishing measures. Rewards were linked to the measures and the new scorecard system was introduced and explained to the couple's six children. The next morning, the family's six-year-old son arose early (at six o'clock) to clean the bathroom—all on his own initiative. From that point on, the house stayed clean. Your takeaway: Metrics motivate.

- At the turn of the millennium, A. G. Lafley, P&G's CEO, lamented that fewer than 10 percent of all new product ideas came from outside the company. Seeking to amp up innovation and drive growth, Lafley announced the new goal: "I'm a big believer that we sometimes need help in solving problems. So I have set a goal to get half of our innovation from outside."[52] A new measure—product ideas generated from outside the organization—became a key talking point. It also provided the motivation to build new organizational capabilities.

  Specifically, with the goal articulated and the metric established, P&G leaders knew they needed to invest in a new vehicle to make it easy for outsiders to share their best ideas. Connect & Develop, an Internet-based crowdsourcing initiative, was put in place. Lafley also realized that to succeed, P&G needed to establish a trust-based collaborative routine: "We want to be known as the company that collaborates—inside and out—better than any other company in the world."[53] By 2010, P&G obtained over 42 percent of its new product ideas from suppliers, customers, and even competitors. Lafley explained,

  > But my point of view was, wherever they come from, you've got to get the people with the idea, the technology, and the ability to execute the idea in the marketplace together. For example, one product is the Swiffer—the technology in the duster that helps it pick up so much debris came from a competitor in Japan called Unicharm.[54]

  Your takeaway: Metrics guide capability development—they can make the improbable possible.

These two examples illustrate the power of measurement. Measurement communicates ideals and motivates behavior. Unfortunately, far too many companies have failed to see that measurement is an essential part of strategy. The reality is that measurement is essential to strategy execution. It also guides strategy development. Without aligned and effective measure, most of what we have talked about in the first five chapters of this book will fail to deliver results. With this in mind, let us tweak and then answer David's question from the opening story: "What are the characteristics of a best-in-class measurement system?" At a minimum, your measurement system should do the following:

- Connect your strategy and your daily efforts to what customers really value. It should help you understand exactly how they are evaluating your performance.

- Communicate expectations and motivate the right behavior—everywhere! You need to make sure your employees, your suppliers, and your customers know where they stand. Celebrate with them when they perform, correct them when they fail, and encourage them to learn always.

- Balance financial and cost measures with important, hard-to-quantify strategic measures that support longer-term strategic initiatives. Capability and relationships development requires supportive measurement.

- Support functional and operational excellence even as it promotes value co-creation across organizational boundaries—within the firm and across the supply chain.

- Help you understand and constantly improve/innovate value-added processes across functional and organizational boundaries.

The bottom line: If you don't measure well, you will never measure up!

Endnotes

1. Collins, J. 2002. *Good to Great.* New York: HarperCollins.
2. Ibid, 1
3. Gates, B. 2013. "My Plan to Fix the World's Biggest Problems." *The Wall Street Journal*, January 25.
4. Cacciola. 2012. "Cracking the Long-Jump Code." *The Wall Street Journal*, February 14.
5. Hayes, R., and Wheelwright, S. 1984. *Restoring Our Competitive Edge: Competing through Manufacturing.* New York: John Wiley and Sons.

6. Haas, E. 1987. "Breakthrough Manufacturing." *Harvard Business Review* 65(2):75–81.

7. Ford Motor Company. 1982. Commercial-Ford-Quality Is Job 1. Retrieved September 28, 2013, from http://www.youtube.com/watch?v=UEB6l6YUx7E

8. Kerr, S. 1975. "On the Folly of Rewarding A, while Hoping for B." *Academy of Management Journal* 18(4):769–783.

9. Fawcett, S., Andraski, J., Fawcett, A., and Magnan, G. 2009. "The Art of Supply Change Management." *Supply Chain Management Review* 13(8):18–25.

10. Fawcett, S., Magnan, G., and Williams, A. 2004. "Supply Chain Trust Is within Your Grasp." *Supply Chain Management Review* 8(2):20–26.

11. Bowersox, D., Calantone, R., Clinton, S., Closs, D., Cooper, M., Droge, C., Fawcett, S., Frankel, R., Frayer, D., Morash, E., Rinehart, L., and Schmitz, J. 1995. *World Class Logistics: The Challenge of Managing Continuous Change*. Oak Brook, IL: Council of Logistics Management.

12. Kelvin, W. 1889. *Electrical Units of Measurement* (Vol. 1). London, UK: Macmillan.

13. Fawcett, S., Smith, S., and Cooper, M. 1997. "Strategic Intent, Measurement Capability, and Operational Success: Making the Connection." *International Journal of Physical Distribution & Logistics Management* 27(7):410–421.

14. Fawcett, S., Ellram, L., and Ogden, J. 2007. *Supply Chain Management: From Vision to Implementation*. Upper Saddle River, NJ: Prentice Hall.

15. Kaplan, R. 1991. "New Systems for Measurement and Control." *The Engineering Economist* 36(3):201–218.

16. Rosen, W. 2012. *The Most Powerful Idea in the World*. Chicago, IL: University of Chicago Press.

17. Fine, C. 1998. *Clockspeed*. Reading, MA: Perseus Books.

18. Fawcett, S., Andraski, J., Fawcett, A., and Magnan, G. 2010. The Indispensable Supply Chain Leader. *Supply Chain Management Review* 14(5):22–29.

19. Bowersox, D., Closs, D., and Cooper, M. 2012. *Supply Chain Logistics Management*. New York: McGraw-Hill/Irwin.

20. Kaplan, R. 1984. "Yesterday's Accounting Undermines Production." *Harvard Business Review* 62:95–101; Eccles, R. 1991. "The Performance Measurement Manifesto." *Harvard Business Review* 69(1):131–137; Kaplan, R. S., and Norton, D. P. 1996. *The Balanced Scorecard: Translating Strategy into Action* Boston: Harvard Business School Press.

21. Skinner, W. 1986. "The Productivity Paradox." *Harvard Business Review* 64(4):55–59.

22. Ibid, 12

23. Waller, M., and Fawcett, S. 2012. "The Total Cost Concept of Logistics: One of Many Fundamental Logistics Concepts Begging for Answers." *Journal of Business Logistics* 33(1):1–3.

24. Ibid, 11, Bowersox et al.

25. Ibid, 14

26. Ibid, 14

27. Ellram, L. 1999. *The Role of Supply Management in Target Costing*. Tempe, AZ: Center for Advanced Purchasing Studies.

28. Fawcett, S., Magnan, G., and Ogden, J. 2007. *Achieving World-Class Supply Chain Collaboration: Managing the Transformation*. Tempe Arizona: Institute for Supply Management.

29. Kurt Salmon Associates, Inc. 1993. *Efficient Consumer Response: Enhancing Consumer Value in the Grocery Industry*. Washington, D.C.: Food Marketing Institute.

30. Ibid, 29

31. Dwyer, K. 2008. "Perfect Order Fulfillment: Getting It All Right." *Inbound Logistics,* April 2008. Retrieved September 27, 2013, from http://www.inboundlogistics.com/cms/article/perfect-order-fulfillment-getting-it-all-right/

32. Banker, S. 2010. "The Perfect Order Metric Is Not Sufficient." *Logistics Viewpoints,* January 28. Retrieved September 27, 2013, from http://logisticsviewpoints.com/2010/01/28/the-perfect-order-metric-is-not-sufficient/

33. Ng, S. 2013. "P&G, Big Companies Pinch Suppliers on Payments." *The Wall Street Journal,* April 16. Retrieved September 28, 2013, from http://online.wsj.com/article/SB10001424127887324010704578418361635041842.html?KEYWORDS=serena+ng+big+companies

34. Bowman, R. 2011. "The Perfect Order Isn't So Perfect." *SupplyChainBrain,* November 21. Retrieved September 28, 2013, from http://www.supplychainbrain.com/content/blogs/think-tank/blog/article

35. Ibid, 34

36. Ibid, 34

37. Kaplan, R., and Norton, D. 1992. "The Balanced Scorecard—Measures That Drive Performance." *Harvard Business Review* 70(1):71–79.

38. Kaplan, R., and Norton, D. 1993. "Putting the Balanced Scorecard to Work." *Harvard Business Review* 71(5):134–137.

39. Kaplan, R., and Norton, D. 2007. "Using the Balanced Scorecard as a Strategic Management System." *Harvard Business Review* 85(7/8):150–161.

40. Ibid, 37

41. Kaplan, R., and Norton, D. 2012. "The Balanced Scorecard's 20th Anniversary." *Harvard Business Publishing Newsletter* 14(3).

42. Ibid, 37, 39

43. Ibid, 39

44. Ibid, 38

45. Ibid, 9; Fawcett, S. E., Magnan, G. M., and McCarter, M. W. 2008. "A Three-Stage Implementation Model for Supply Chain Collaboration." *Journal of Business Logistics* 29(1):93–112.

46. Day, M., Fawcett, S. E., Fawcett, A. M., and Magnan, G. M. 2013. "Trust and Relational Embeddedness: Exploring a Paradox of Trust Pattern Development in Key Supplier Relationships." *Industrial Marketing Management* 42(2):152–165; Fawcett, S. E., Jones, S., and Fawcett, A. M. 2012. "Breakthrough Trust: The Catalyst to Collaborative Innovation." *Business Horizons* 55(2):163–178.

47. Ibid, 28

48. Coyle, J., Langley, C., Gibson, B., and Novack, R. 2011. *Supply Chain Management: A Logistics Perspective*. Florence, KY: South-Western College Publications.

49. Bowersox, D., Closs, D., and Cooper, M. 2012. *Supply Chain Logistics Management*. New York: McGraw-Hill/Irwin.

50. Ibid, 48

51. Lambert, D., and Stock, J. 1982. "Using Advanced Order Processing Systems to Improve Profitability." *Business Horizons* 32(3):23–29.

52. Florian, E., Smith, F., Barrett, C., Lafley, A., McNerney, J., and Collins, A. 2004. "Special: CEOs on Innovation." *Fortune,* March 8. Retrieved September 28, 2013, from http://money.cnn.com/magazines/fortune/fortune_archive/2004/03/08/363668/index.htm

53. Huston, L., and Sakkab, N. 2006. "Connect and Develop: Inside Procter & Gamble's New Model for Innovation." *Harvard Business Review* 84(3):58–66.

54. Ibid, 52

Index

Numbers
3M, knowledge (IT enablement hierarchy), 142-143
80/20 rule (Pareto principle), 77-78

A
ABC (activity-based costing), 83-85
ABC classification, 77-82
acquisition touch points, 24
activity-based costing (ABC), 83-85
affective processes, expectations and, 16-17
alliance development process, 69-74
    collaborative planning phase, 72-73
    day-to-day management phase, 73-74
    internal planning phase, 72
Amazon.com, supply chain configuration, 97-98
Apple, product innovation, 12
appropriateness in customer relationship management, 74-75
assessment. *See* measuring
automation, as cost reduction strategy, 9
availability in order fulfillment, 40-41

B
B2B (business to business), 7
back orders, cost of, 54
balanced scorecards, 188-194
bar codes, explained, 152
behavior motivation, 176-177
Best Buy, showrooming, 6
BI (business intelligence), explained, 152
Bob Evans, collaborative capability, 75
Boeing, supply chain configuration, 120
business intelligence (BI), explained, 152

C
cash-to-cash cycle time, 186-187
chain reaction, order fulfillment as, 95-97
change, difficulty of, 112-115
cognitive processes, expectations and, 16-17
collaborative planning phase (alliance development), 72-73
communication gaps, 26
comparison shopping, 6
compatibility in global network design, 119-120
competitive advantage, 27-29
complexity
    logistics system, 107-108
        operating network, 102-103
        organizational structure, 101-102
        SKU management, 103-105
        supply base, 105-106
        value-added processes, 102
    sources of, 101-108
        customer base, 106-107
    of supply chain configuration, 99-108
configuration
    in global network design, 120
    of supply chain, 97-115
        complexity of, 99-108
        difficulty of change, 112-115
        global implications, 118-122
        systems thinking, 115-118
        tradeoffs, 108-112
connectivity in information sharing, 138-140
consistency of delivery, 42
constraints in order fulfillment, 117
contingency plans in order fulfillment, 43-44
continuity planning in global network design, 121-122
control in global network design, 121
coordination in global network design, 120-121

cost of failure
    "dropping the baton," 63-65
    order fulfillment, 54-57
        out-of-stock occurrences, 54-57
        supply chain glitches, 57
cost reduction
    in customer value, 9
    logistics and, 15
CRM (customer relationship management)
    components of, 155-158
    in customer fulfillment strategy, 66-67
        key points, 74-77
        strategic alliances, 68-74
        transactional relationships, 67-68
    drawbacks of, 159
    explained, 152, 155-159
cross-functional processes, 109-112
customer base, as source of complexity, 106-107
customer delight, 21
customer fulfillment strategy
    cost of "dropping the baton," 63-65
    customer relationship management in, 66-67
        key points, 74-77
        strategic alliances, 69-74
        transactional relationships, 67-68
    developing, 66
    segmentation tools
        ABC classification, 77-82
        customer profitability analysis, 82-89
    service offerings in, 89-91
customer profitability analysis, 82-89
customer relationship management (CRM)
    components of, 155-158
    in customer fulfillment strategy, 66-67
        key points, 74-77
        strategic alliances, 68-74
        transactional relationships, 67-68
    drawbacks of, 159
    explained, 152, 155-159
customer satisfaction strategies, 17-21
    limitations, 18
    mission statements, 20
customer service strategies, 17
    limitations, 18
    mission statements, 20

customer success strategies, 21-22
    limitations, 18
    mission statements, 20
customer-centric measurement, 187-188
customer-experience systems. *See* service system design
customers
    of choice, service offerings for, 89-90
    determining needs of, 4-6
    identifying, 3-4
    satisfaction
        customer satisfaction strategies, 17-21
        customer service strategies, 17
        customer success strategies, 21-22
        result of expectations, 15-17
    service system design, 23-29
        loyalty and competitive advantage, 27-29
        orchestration, 24-25
        touch points, 24
        value gaps, 25-26
    technology empowerment, 6-7
    value creation for, 7-15
        cost reduction, 9
        delivery capability, 10-11
        logistics, 14-15
        process innovation, 12-13
        quality, 9-10
        responsiveness, 11-12
        total performance, 13-14

D

data (IT enablement hierarchy), 142, 145-146
data analytics in CRM systems, 157-158
data display in CRM systems, 158
data storage in CRM systems, 156
data-capture technology in CRM systems, 155-156
day-to-day management phase (alliance development), 73-74
delight (customers), 21
deliverables of order fulfillment systems, 39-45
    efficient operations, 45
    product availability, 40-41
    service recovery, 43-44
    timely delivery, 41-43
    transparency, reliability, 43

delivery capability
    in customer value, 10-11
    logistics and, 15
    SCOR delivery process model, 46-51
        consolidate orders, 48-49
        install product, 51
        invoice, 51
        load product/generate shipping documents, 50
        pack product, 50
        pick product, 50
        plan/build loads, 49
        process inquiry and quote, 47
        receive product from source, 49
        receive/enter/validate order, 48
        receive/verify product by customer, 50-51
        reserve inventory/determine delivery date, 48
        route shipments, 49
        select carriers, 49
        ship product, 50
Dell, process innovation, 12
difficulty of change, 112-115
"dropping the baton," cost of, 63-65
Drucker, Peter, 3

E

EDI document standard, explained, 152
efficient operations in order fulfillment, 45
80/20 rule (Pareto principle), 77-78
Einstein, Albert, 174
Eliot, T.S., 134
empathy in customer relationship management, 77
enabling technology. *See* technology
evaluations
    by customers, 6-7
    of order fulfillment, 35-37
execution, performance measurement and, 177-178
expectations, customer satisfaction and, 15-17

F

failure cost in order fulfillment, 54-57
    out-of-stock occurrences, 54-57
    supply chain glitches, 57

fill rate, 41
    measuring, 196-198
flexibility
    in customer value, 11-12
    of delivery, 42-43
    logistics and, 15
Ford, rewards and performance measurement, 174
form utility, 7
future opportunities in customer relationship management, 75

G

global network design
    as cost reduction strategy, 9
    supply chain configuration and, 118-122
GM, rewards and performance measurement, 174
goals
    in balanced scorecards, 191-192
    competing versus complementary, 115-116

H

high potential customers, service offerings for, 90

I

information (IT enablement hierarchy), 142-143
information sharing with technology, 136-141
    connectivity, 138-140
    willingness, 140-141
information technology (IT). *See* IT (information technology)
innovation
    logistics and, 15
    process innovation, 12-13
internal planning phase (alliance development), 72
investment patterns, as obstacle to IT enablement, 142-145
invoice processing, 51
IT (information technology)
    information sharing, 136-141
        connectivity, 138-140
        willingness, 140-141

obstacles to IT enablement, 142-149
    failure to follow proven path, 145-149
    investment patterns, 142-145
role of, 131-133
service systems, 149-162
    CRM (customer relationship management) systems, 155-159
    defined, 149-150
    IT-enabling tools, 151
    order processing systems, 159-162
strategy implementation, 134-136

K

Kelvin, Lord, 66
key accounts, service offerings for, 89-90
knowledge (IT enablement hierarchy), 142-143
knowledge gaps, 26
Kroger, data analytics, 157

L

L. L. Bean, customer service policy, 68
lifetime customer value, measuring, 76-77
logistics
    in customer value, 14-15
    as source of complexity, 107-108
lost customers, cost of, 56
lost sales, cost of, 54-56
loyalty, 27-29

M

measuring performance, 148, 171-174
    balanced scorecards, 188-194
    behavior motivation, 176-177
    customer-centric measurement, 187-188
    execution resulting from, 177-178
    importance of, 175-178
    performance versus profitability tradeoffs, 194-201
    rewards and, 174-175
    supply chain measurements, 183-187
    total costing method, 180-182
    traditional measurements, 178-180
    understanding provided by, 176

mission statements, 20
moment of truth, 24
motivation from performance measurement, 176-177

N

National Semiconductor, systems thinking, 118
needs of customers, determining, 4-6
network. *See* customer fulfillment strategy; order fulfillment; supply chain

O

on-time delivery in order fulfillment, 41-43
operating network, as source of complexity, 102-103
operational efficiency in order fulfillment, 45
opportunity analysis, 72
orchestration, 24-25
order cycle, 45-46
    measuring cycle time, 198-201
    SCOR delivery process model, 46-51
        consolidate orders, 48-49
        install product, 51
        invoice, 51
        load product/generate shipping documents, 50
        pack product, 50
        pick product, 50, 86
        plan/build loads, 49
        process inquiry and quote, 47
        receive product from source, 49
        receive/enter/validate order, 48
        receive/verify product by customer, 50-51
        reserve inventory/determine delivery date, 48
        route shipments, 49
        select carriers, 49
        ship product, 50
order fulfillment. *See also* customer fulfillment strategy
    as chain reaction, 95-97
    cost of failure, 54-57
        out-of-stock occurrences, 54-57
        supply chain glitches, 57

deliverables of, 39-45
   efficient operations, 45
   product availability, 40-41
   service recovery, 43-44
   timely delivery, 41-43
   transparency, reliability, 43
enabling technology
   information sharing, 136-141
   obstacles to, 142-149
   strategy implementation, 134-136
evaluating, 35-37
order cycle, 45-46
   SCOR delivery process model, 46-51
order processing systems, 159-162
out-of-stock statistics, 38-39
postsales customer service, 52-54
supply chain configuration, 97-115
   complexity of, 99-108
   difficulty of change, 112-115
   global implications, 118-122
   systems thinking, 115-118
   tradeoffs, 108-112
order processing systems, explained, 159-162
orders shipped complete, 41
organizational structure, as source of complexity, 101-102
out-of-stock occurrences
   cost of, 54-57
   statistics on, 38-39
   stockout frequency, 41
outsourcing, as cost reduction strategy, 9

P

Pareto principle, 77-78
PepsiCo, customer success strategies, 22
perception gaps, 26
perfect orders
   defined, 43
   as supply chain measurement, 183-185
performance gaps, 26
performance measurement, 148, 171-174
   balanced scorecards, 188-194
   behavior motivation, 176-177
   customer-centric measurement, 187-188
   execution resulting from, 177-178

   importance of, 175-178
   performance versus profitability tradeoffs, 194-201
      order cycle time, 198-201
      product availability, 196-198
   rewards and, 174-175
   supply chain measurements, 183-187
   total costing method, 180-182
   traditional measurements, 178-180
   understanding provided by, 176
pick-to-light systems, explained, 152
place utility, 8
possession utility, 8
postsales customer service, 52-54
process innovation
   in customer value, 12-13
   technology and, 134-136, 148
product availability
   measuring, 196-198
   in order fulfillment, 40-41
product support, postsales customer service, 52-53
productivity enhancement, as cost reduction strategy, 9
profitable growth
   customer profitability analysis, 82-89
   in customer relationship management, 75-76
   performance versus profitability tradeoffs, 194-201
      order cycle time, 198-201
      product availability, 196-198
proven path to IT enablement, 145-149

Q

quality
   in customer value, 9-10
   logistics and, 15

R

reliability in order fulfillment, 43
responsiveness
   in customer value, 11-12
   logistics and, 15

Retail Link (Walmart), 4
retention of customers in customer relationship management, 75-76
returns, postsales customer service, 53-54
reviews by customers, 6-7
rewards, performance measurement and, 174-175
RFID, 139, 152

S

SAMBC (Service As Measured By the Customer), 187-188
satisfaction
    customer satisfaction strategies, 17-21
    customer service strategies, 17
    customer success strategies, 21-22
    result of expectations, 15-17
satisfaction gaps, 26
SCOR delivery process model, 46-51
    consolidate orders, 48-49
    install product, 51
    invoice, 51
    load product/generate shipping documents, 50
    pack product, 50
    pick product, 50
    plan/build loads, 49
    process inquiry and quote, 47
    receive product from source, 49
    receive/enter/validate order, 48
    receive/verify product by customer, 50-51
    reserve inventory/determine delivery date, 48
    route shipments, 49
    select carriers, 49
    ship product, 50
segmentation of customers. *See* customer fulfillment strategy
segmentation tools
    ABC classification, 77-82
    customer profitability analysis, 82-89
Service As Measured By the Customer (SAMBC), 187-188
service delivery gaps, 26
service offerings in customer fulfillment strategy, 89-91
service quality gaps, 26
service recovery in order fulfillment, 43-44

service system design, 23-29
    loyalty and competitive advantage, 27-29
    orchestration, 24-25
    systems thinking, 115-118
    technology and, 149-162
        CRM (customer relationship management) systems, 155-159
        defined, 149-150
        IT-enabling tools, 152
        order processing systems, 159-162
    touch points, 24
    value gaps, 25-26
showrooming, 6
Six Sigma, 10
SKU management, as source of complexity, 103-105
Sony de Mexico
    order fulfillment, 36-37
    systems thinking, 118
spare parts, postsales customer service, 53
specification gaps, 26
speed of delivery, 42
stock-keeping units. *See* SKU management
stockouts. *See* out-of-stock occurrences
strategic alliances, managing, 69-74
    collaborative planning phase, 72-73
    day-to-day management phase, 73-74
    internal planning phase, 72
success of customers. *See* customer success strategies
superordinate goal, establishing, 115-116
supply base, as source of complexity, 105-106
supply chain
    configuration, 97-115
        complexity of, 99-108
        difficulty of change, 112-115
        global implications, 118-122
        systems thinking, 115-118
        tradeoffs, 108-112
    cost of glitches, 57
    customer value, 7-15
        cost reduction, 9
        delivery capability, 10-11
        logistics, 14-15
        process innovation, 12-13
        quality, 9-10

responsiveness, 11-12
total performance, 13-14
customers
determining needs of, 4-6
identifying, 3-4
order cycle, 45-46
SCOR delivery process model, 46-51
performance measurements, 183-187
on-time delivery, 41-43
total costing, 180-182
system boundaries, defining, 116
systems thinking in order fulfillment, 115-118

T

tailored logistics. *See* customer fulfillment strategy
tailored metrics in balanced scorecards, 192-193
Target, supply chain configuration, 98
technical information, postsales customer service, 52
technology
information sharing, 136-141
connectivity, 138-140
willingness, 140-141
obstacles to IT enablement, 142-149
failure to follow proven path, 145-149
investment patterns, 142-145
role of, 131-133
service systems, 149-162
CRM (customer relationship management) systems, 155-159
defined, 149-150
IT-enabling tools, 152
order processing systems, 159-162
strategy implementation, 134-136
technology empowerment of customers, 6-7
Tesco, data analytics, 157-158
3M, knowledge (IT enablement hierarchy), 142-143
time utility, 8
timely delivery in order fulfillment, 41-43
TMS (transportation management system), explained, 152
total costing method, 180-182
total performance in customer value, 13-14

touch points, 24
Toyota
responsiveness, 12
rewards and performance measurement, 174
on-time delivery, 41
tradeoffs
evaluating, 117
performance versus profitability, 194-201
order cycle time, 198-201
product availability, 196-198
in supply chain configuration, 108-112
traditional performance measurements, 178-180
transactional relationships
managing, 67-68
service offerings for, 90
translation gaps, 26
transparency in order fulfillment, 43
transportation management system (TMS), explained, 152

U

understanding, acquiring from performance measurement, 176
United Airlines, complaints against, 7
utilities, list of, 7-8
utilization touch points, 24

V

value
creating for customers, 7-15
cost reduction, 9
delivery capability, 10-11
logistics, 14-15
process innovation, 12-13
quality, 9-10
responsiveness, 11-12
total performance, 13-14
lifetime customer value, measuring, 76-77
utilities, 7-8
value gaps, 25-26
value propositions, 23
value-added processes, as source of complexity, 102

W

Walmart
    customers, determining needs of, 4-6
    process innovation, 12-13
    supply chain configuration, 98-115
        complexity of, 99-108
        difficulty of change, 112-115
        global implications, 118-122
        tradeoffs, 108-112
    wisdom (IT enablement hierarchy), 143-144
warehouse management system (WMS),
    explained, 152
warranty, postsales customer service, 53
what-if analysis, 73
willingness in information sharing, 140-141
wisdom (IT enablement hierarchy), 143-144
WMS (warehouse management system),
    explained, 152

X

Xerox, supply base management, 105-106
[x+1] Inc., CRM (customer relationship
    management), 155